Renaissance M

*For Ed Ricketts Jr., who provided support
and encouragement and became a great friend*

Renaissance Man of Cannery Row

The Life and Letters of Edward F. Ricketts

Edited and with an Introduction by
Katharine A. Rodger

The University of Alabama Press
Tuscaloosa and London

Copyright © 2002
The University of Alabama Press
Tuscaloosa, Alabama 35487-0380
All rights reserved
Manufactured in the United States of America

Unless otherwise noted, all photographs courtesy of Edward F. Ricketts Jr.

Typeface: New Baskerville and Voluta Script

∞

The paper on which this book is printed meets the minimum requirements of American National Standard for Information Science-Permanence of Paper for Printed Library Materials, ANSI Z39.48-1984.

Library of Congress Cataloging-in-Publication Data

Ricketts, Edward Flanders, 1896–1948.
Renaissance man of cannery row : the life and letters of Edward F. Ricketts / edited by Katharine A. Rodger.
p. cm.
Includes bibliographical references and index.
ISBN 0-8173-1172-6 (alk. paper)
1. Ricketts, Edward Flanders, 1896–1948. 2. Marine biologists—West (U.S.)—Biography. 3. Steinbeck, John, 1902–1968—Friends and associates. I. Rodger, Katharine A. (Katharine Anne), 1974– II. Title.
QH31.R53 A3 2002
578.77′092—dc21

2002002308

British Library Cataloguing-in-Publication Data available

Contents

Acknowledgments
vii

Editor's Note
ix

Introduction
xi

Biographical Essay
xv

1936–1938
1

1939–1940
36

1941–1942
100

1943–1945
176

1946–1948
219

Works Cited
275

Index
279

Illustrations follow page 84

Acknowledgments

I would first and foremost like to thank Susan Shillinglaw for her support during the past two years. She has been an outstanding mentor, thesis advisor, and friend to me, and her dedication to this project has been tireless. Professors Robert Cullen, Donald Keesey, and Scott Rice, also at San José State University, readily provided support throughout my graduate program. I am grateful for the outstanding resources made available to me at San José State University, particularly at the Center for Steinbeck Studies. Polly Armstrong and Maggie Kimball were of invaluable assistance during the course of my research at Stanford University's Department of Special Collections. I thank Eugene Winick at McIntosh and Otis for his guidance regarding permissions. I also thank the Joseph Campbell Foundation for granting permission to include excerpts from Joseph Campbell's letters.

This book would not have been possible without the support and contributions of the Ricketts family, particularly Ed, Nancy, and Lisa. Also, thank you to Donald Keith Henry for careful proofreading.

My family—Amy, David, and Kathy Rodger—have been a constant source of support and love, and I would be truly lost without them. Thank you to the friends and fellow graduate students who have read drafts, listened to me read letters aloud, and patiently allowed me to talk incessantly about a man many have never before heard of. I love you all.

And my sincerest gratitude to Mr. Jim Levitt, who generously provided funding for my work and whose enthusiasm and belief in the importance of this project continue to inspire me.

Editor's Note

Most of the 136 selected letters in this volume have been transcribed from carbon copies in the Edward F. Ricketts collection—which includes approximately 300 letters—housed in Special Collections at Stanford University. The letters reveal the scope of Ricketts's interests and the many diverse people he corresponded with during the last twelve years of his life.

According to those who knew him, Ricketts's writing style reflects the way he thought and spoke; he admired good writing and often spent a great deal of time—even in composing his personal correspondence—trying to clearly express his ideas. As a result, many letters have a colloquial quality that conveys Ricketts's personality, particularly his sense of humor.

In transcribing these letters, small errors and inconsistencies in Ricketts's spelling, grammar, and punctuation have been corrected without comment. Overall, however, his prose has not been altered unless indicated. Ricketts typed his correspondence with carbon paper in order to copy each letter for his files. Some carbon copies do not bear his signature; those letters remain unsigned in this volume. Often, Ricketts abbreviates names, words, and places. Abbreviations he habitually used have not been changed or noted within the text of the letters themselves, although any that hinder coherence have been clarified in brackets. The following is a list of Ricketts's most common abbreviations:

acct	on account of
altho	although
Bkly	Berkeley, California
BPT	*Between Pacific Tides*
Crl	Carol Steinbeck
Crml	Carmel, California
Dk and Jn	Dick and Jan Albee

Ed Jr, Edw Jr	Edward F. Ricketts Jr.
etc	etc.
FF, F/F	Fred and Frances Strong
Jon, Jn, Sbeck	John Steinbeck
MB	Monterey Bay
non-tel	nonteleological
PBL	Pacific Biological Laboratories
PG	Pacific Grove, California
R/T	Ritch and Tal Lovejoy
SF	San Francisco, California
T	Toni Jackson
tho	though
thot	thought
thru	through
UC	University of California, Berkeley
VG	very good
VVG	very, very good

Introduction

Edward Flanders Robb Ricketts is perhaps most widely known as the model for Doc in John Steinbeck's novella *Cannery Row* (1945): a marine biologist who drinks beer milkshakes, loves women, hates getting his head wet, pays bums to collect frogs, and ends parties with a reading of the anonymous eleventh-century Sanskrit poem "Black Marigolds." He might have also been a model for Slim in *Of Mice and Men* (1937), Doc Burton in *In Dubious Battle* (1936), Jim Casy in *The Grapes of Wrath* (1939), Dr. Winter in *The Moon Is Down* (1942), and Lee in *East of Eden* (1952). He first appeared in fiction as Doctor Phillips in a much anthologized short story "The Snake" (1935). Steinbeck's biographer, Jackson Benson, calls this far-seeing character Steinbeck's "Merlin figure"—the wise seer, the man of vision. In short, Ed Ricketts, Steinbeck's closest friend throughout the 1930s and 1940s, was also a kind of alter ego for the writer. "His mind had no horizons. He was interested in everything," writes Steinbeck in "About Ed Ricketts" (1951), a sensitive portrait of his friend that reveals Ricketts's humor, intellect, and humanity.

As elaborate as Steinbeck's fictional portraits and impressions are, Ricketts the man is a more respected scientist, complex thinker, and valued friend than even Steinbeck's depictions reveal him to be. As Richard Astro clarifies in his seminal study *John Steinbeck and Edward F. Ricketts: The Shaping of a Novelist* (1973), "Edward F. Ricketts was a highly complicated individual [and] a devoted and highly rational biologist who sought to uncover scientific truths" (25). He read and wrote extensively about biology, ecology, music, poetry, and religion. People were drawn to Ricketts, and he befriended numerous prominent scientists, artists, and scholars throughout his life, many of whom he corresponded with regularly. After his death in 1948, he left a large volume of correspondence from the final twelve years of his life. (A devastating fire in 1936 destroyed his earlier letters and papers.) Ricketts's surviving letters—which, along with notebooks and essays, are housed in Special Collections at Stanford University—reveal the wide spectrum of his interests and achievements during the most productive years of his life.

As a marine biologist, Ricketts was something of an innovator and pioneer. His boundless curiosity and keen interest in observing and recording species of marine animals were maintained throughout his life. From 1923 to 1948, he owned and operated Pacific Biological Laboratories on Cannery Row in Monterey, which supplied marine specimens to schools and laboratories around the country. Ricketts won the respect of fellow scientists who worked nearby at Stanford's Hopkins Marine Station as well as of marine biologists around the world. Although Stanford University initially balked at publishing Ricketts's *Between Pacific Tides*, a handbook on Pacific marine life, because he did not have a university degree, the press eventually published the study in 1939, and now it is in its fifth edition. Ricketts made a name for himself in scientific circles because of his commitments to investigating marine life, to developing an ecological, holistic perspective, and to mapping the Pacific littoral throughout his career. He also studied and wrote extensively in the 1940s on the disappearance of the sardines in Monterey Bay. Ricketts produced data and documents about plankton populations, environmental conservation, and wave shock that continue to interest scholars and students. His "comments on science, the scientific process, and the study of nature," notes nuclear chemist Peter A. J. Englert, "are highly relevant to the education of today's environmental scientists" (177).

Although Ricketts went on several collecting trips along the Pacific coast, his most famous expedition was the one he and Steinbeck took in 1940 to the Gulf of California. The trip resulted in *Sea of Cortez* (1941), one of Steinbeck's most complex statements about his environmentalism and philosophy. While many critics assume Steinbeck's contribution to *Sea of Cortez* was the log portion and Ricketts's the phyletic catalogue, the log was actually based largely on Ricketts's notebook. Focusing on the experience of their trip and not merely on facts—the "toto-picture" is the phrase Ricketts often uses—the men state their intention to "deal with life, with teeming boisterous life, and learn something from it, learn that the first rule of life is living" (29). The resulting text integrates science, travel, and philosophy and is recognized today as one of the century's seminal works on ecological holism. In 1951, Viking Press published *The Log from the* Sea of Cortez, which consisted of the narrative of the log and Steinbeck's essay "About Ed Ricketts"; the phyletic catalogue was not included.

Ricketts had a wide range of interests and read extensively on philosophy, poetry, and music theory. "Ed's interest in music was passionate and profound," Steinbeck recalls in "About Ed Ricketts." "He thought of it as deeply akin to creative mathematics" (254). Hanging on the walls of the lab were charts mapping the finest composers in history, including Bach, whose *Art of the Fugue* Ricketts believed to be one of the greatest pieces ever composed. He created similar outlines, on a smaller scale, of the world's poets, authors, and painters—outlines he often alludes to in his letters. Eastern thinkers like Lao-tzu (credited as the author of the *Tao-te Ching*, the cornerstone text of Taoism) and Krishnamurti (an influential spiritual teacher from India) were among the most enlightened in Ricketts's estimation, and he saw Krishnamurti whenever he visited the Monterey area. Also, Ricketts read the poetry of Walt Whitman, Robinson Jeffers, Li Po, and John Keats, and "enlarged his scientific German so that he could read *Faust*," as Steinbeck notes. "Ed's mind seems to me to have been a timeless mind, not modern and not ancient" (255).

Although Ricketts was perhaps most revered among his friends as a talker—some called him the Buddha because his words were wise and because he could listen attentively—he also explored his ideas through writing—in letters, notebooks, and essays. His curiosity was boundless, and he left behind an abundance of written material containing his research, theories, and meditations about ideas ranging from art and literature, including James Joyce's *Finnegans Wake*, to the medicinal attributes of sharks' liver oil. In the last two decades of his life he worked hard on revising three essays: one on ecological holism, another on poetry, and a third on transcendence, which he called "breaking through." References to these essays and the ideas therein appear throughout Ricketts's letters, along with details about collecting trips, family news, and local gossip.

Known to be a keen listener, Ricketts was often sought out for advice. He met an amazing number of people during his life, including artists and writers such as mythologist Joseph Campbell, author Henry Miller, local painter Ellwood Graham, and watercolorist James Fitzgerald. This volume of selected correspondence attests to the great number of friendships he maintained. In many respects, Cannery Row was a haven of Bohemian culture in the 1930s and 1940s, and Ricketts's lab was a center where a diverse group gathered to talk, drink, and, on occasion, have long and boisterous parties—parties Steinbeck recalls in both *Cannery Row* and *Sweet Thursday* (1954). Ricketts's own interests in art, music, literature, and philosophy often ignited discussions that extended beyond the lab and into lifelong correspondence with friends and colleagues. "Ed carried on a large and varied correspondence with a number of people," Steinbeck recalls. "He answered letters quickly and at length, using a typewriter with elite type to save space" ("About Ed Ricketts" 231).

Ricketts's letters from 1936 to 1948 are a record of people and interests most dear to him. He makes repeated reference to his three philosophical essays, which circulated among friends and colleagues for criticism and revision; to marital issues concerning his wife, Nan, from whom he was permanently separated in 1936; to relationships of friends; to collecting trips and plans for future projects; and to his three children's activities and achievements. Significantly, his letters also document events that transpired in his friends' lives, particularly Steinbeck's, as he was the author's closest friend and confidant throughout the 1930s and 1940s. Ricketts expresses his concerns, surprises, and opinions about Steinbeck's life and work, and his letters reveal the fascinating and complex relationship they shared. Most importantly, however, an intimate understanding of Ricketts's individuality and uniqueness emerges not through the fictional portraits rendered by Steinbeck, but through Ricketts's own words.

Biographical Essay

1897–1930: From Chicago to California

Edward Flanders Robb Ricketts was born on May 14, 1897, in Chicago, Illinois, to Abbott Ricketts and Alice Beverly Flanders Ricketts. His father was a native of Owensville, Kentucky, and his mother, of Haverhill, Massachusetts. Frances Strong, Ricketts's younger sister, kept a journal about their family, revealing that most of their paternal relatives were ministers, while many in their mother's family were storekeepers. "Not a really poor person on either side of the family—as far as I can tell," she writes in her notebook. "Wonder what is the matter with our branch of the family in this generation?" (Strong 1). The family lived on the modest income Abbot Ricketts made as a salesman.

As a child, Ed Ricketts did not excel in sports or physical activities but "was, from birth, a child of intelligence and rare charm. [. . .] He began speaking very young and was using whole but simple sentences before he was a year old" (Strong 1). He was a good child and a protective older brother to his sister, Frances, born in 1899, and brother, Thayer, born in 1902. Their mother, Alice, who worried about their welfare in a rough section of Chicago, kept them relatively sheltered in their neighborhood as children. "We spent [many] hours at home in pre-school days with our noses pressed against the window pane looking out," Frances writes. "Some passersby were daily important events to us" (2). All became avid readers. Ed Ricketts recalls in a letter to Harcourt, Brace and Co., "At the age of six, I was ruined for any ordinary activities when an uncle who should have known better gave me some natural history curios and an old zoology textbook. Here I saw for the first time those magic and incorrect words 'coral insects'" (August 31, 1942). And so began his fascination with the "little animals," as he called them, that continued throughout his life.

At the age of seven, Ricketts began attending Ryerson Public School. Frances reminisces:

There were no serious problems in school for Ed. The teachers always liked him in spite of the fact that he occasionally, when they were too wrong, kindly and politely corrected them. And, as perhaps isn't usual with boys who always get top grades, he was well liked by the boys, and as far as I know, always by the girls. Altho [sic] his buddies sometimes called him "the walking dictionary" he wasn't a bookworm. He didn't take any interest even in sports or athletics, but was strong and was never thought of at all as a sissy. He was always a little short—his mature height was only 5'7". He was compactly and sturdily built, without any fat. He was always pale, having the sort of skin that hardly changed color even with much time in the sun.

During school days he "hardened himself" by taking cold plunges every morning and exercising at night with a wire spring arrangement that he fastened inside the door in his bedroom. By the time he was 11 or 12 he also slept outdoors on the ground in our back yard much of the time rolled in blankets, without "even a tent," until winter. This annoyed and delighted all his boy friends but worried our parents a great deal. It was part of his program to sleep out even during storms. Our parents were pleased when they were able to bribe him to sleep indoors during the coldest weather. (Strong 4)

Even as a child, Ed's intelligence and individuality were apparent to his family and friends.

In 1907, when Ricketts was ten years old, the family moved to Mitchell, South Dakota, because Abbott Ricketts accepted a job as a traveling salesman-auditor. Alice, however, was not happy living in the small town, so the family stayed for only one year. But it was an important year for Ed. His interest in wildlife flourished in this new setting, and he collected and studied birds, insects, and every other form of life he encountered. All of the Ricketts children enjoyed this rural environment, taking advantage of experiences different from those in the city. In her notebook, Frances fondly remembers one instance in which she and her brothers joined a group of children playing in a city block flooded with rainwater. The Ricketts children had not noticed their father among the adult spectators. "Being a most kind father," he let his children continue their games and "went on home to prepare [their] mother for the shock." When the three came home, they found their parents "waiting for [them] with quinine capsules (to ward off colds and pneumonia), hot towels, dry pajamas and warm blankets, and no penalties, not even a scolding" (Strong 5). Perhaps the understanding and accepting nature Ricketts himself was known for in his adult life was derived, at least in part, from the example set by his parents.

In 1908 the Rickettses moved back to Chicago in time for Ed to begin attending high school, where he excelled in both the sciences and the humanities. He enjoyed most of his studies, often drawing connections among different disciplines, a habit of mind that later made his approach to biology "as much philosophical as it was scientific" (Benson 187). Whitman, whose poetry Ricketts considered some of the greatest verse ever written, was one of Ed's favorite poets as an adolescent. His lifelong love of poetry is apparent in his later essays and letters, which include passages from Goethe, Jeffers, and Keats, among others.

After graduating from high school, Ricketts enrolled at Illinois State Normal University. He stayed for one year (1915–16) before dropping out to explore the country. Traveling on his own, he worked in El Paso, Texas, as a bookkeeper in a country club and as a surveyor's assistant in New Mexico. Soon thereafter, in September 1917, he was drafted into World War I as a clerk in the Medical Corps at Camp Grant in Illinois. In March 1919, after the armistice, Ricketts was discharged, having served less than one year. While little is known about his tour in the war, he later reflected upon his superior officer in one of his journals. "During World War One I had one consuming desire: to kill my top sergeant. And this choice hatred lasted for months after I had received my discharge from the army. He seemed to me then an unnecessarily brutal man. Now I understand better. He may have been merely one of those many whose latent sadism push them aggressively into superior positions in the handling of men" ("New Series, No. 2"). In this, as always, Ricketts connects, or integrates, his experiences in an effort to bring humor, perspective, and acceptance into his life. In "About Ed Ricketts," Steinbeck comments on Ricketts's military tours. "In appearance and temperament Ed was a remarkably unmilitary man, [. . .] one would have thought that his complete individuality and his uniqueness of approach to all problems would have caused him to go crazy in the organized mediocrity of the Army. Actually the exact opposite was true. He was a successful soldier. In spite of itself, the Army—at least that part of it which sheltered him—was gradually warped in his favor and for his comfort. He was quite happy in the Army in both wars" (249). Another, and quite humorous, example of Ricketts's acceptance is found in a letter he wrote during World War II to the Ingersoll-Waterbury Company after receiving a defective watch as a Christmas gift. "I say if a watch won't run at all it's no good; it's just simply no god dam good at all Mr. Ingersoll, no matter if you did make it or if it's very pretty G-I khaki-colored and looks swank as hell on your wrist or on anybody's wrist I don't care who. And now I think my wife is pregnant though you can't blame the watch for that, directly, anyhow" (May 8, 1943). While Ricketts's sense of fun is quite apparent, Richard Astro emphasizes that "his is a philosophy of understanding and acceptance in which he seeks to unify experience, to relate the unrelatable so that even nonsense wears a crown of meaning" (*Steinbeck and Ricketts* 28).

Upon his discharge from World War I, Ricketts enrolled at the University of Chicago in the summer of 1919. After six months of living with his family and attending school full time—studying philosophy and Spanish in addition to biology—Ricketts and two roommates moved into an apartment on the south side of the city. He worked at the Sinclair Refining Company and limited his studies to part time, maintaining consistently high grades. After a few months in school, however, Ricketts—like Steinbeck himself, who dropped in and out of Stanford University—became restless and spent two semesters on a walking trip to the southern United States. After traveling to Indianapolis by train from Chicago, he walked through Indiana, Kentucky, North Carolina, and Georgia, talking with and observing all that he could. Steinbeck later describes this trip in his depiction of Doc in *Cannery Row:*

> Once when Doc was at the University of Chicago he had love trouble and he had worked too hard. He thought it would be nice to take a very long walk. He put on a little knapsack and he walked through Indiana and Kentucky and North Carolina and Georgia clear to Florida. He walked among farmers and mountain people, among the swamp people and fishermen. And everywhere people asked him why he was walking through the country.

Because he loved true things he tried to explain. He said he was nervous and besides he wanted to see the country, smell the ground and look at grass and birds and trees, to savor the country, and there was no other way to do it save on foot. And people didn't like him for telling the truth. [. . .] And so he stopped trying to tell the truth. He said he was doing it on a bet—that he stood to win a hundred dollars. (103–04)

Steinbeck biographer Jackson Benson notes that Ricketts did, in fact, tell people he was attempting to win a bet, and Benson asserts that Ricketts's sister, Frances, confirmed the story (190). In the June 1925 issue of *Travel*, Ricketts published an account of this trip in which he recorded observations about the landscape and people he encountered. He included a reference to John Muir's similar journey, published in 1916: "I came across John Muir's 'Thousand-Mile Walk,' and was very interested in his half-century-old description of the country I had just traversed. His suggestion regarding cemeteries proved particularly acceptable" (48). Muir's suggestion to sleep in cemeteries to avoid unwanted intrusions while camping alone, proved useful to Ricketts, who "thereafter, whenever it was convenient, [. . .] spent the night in a 'City of the Dead'" (48).

Returning to the University of Chicago in 1921, Ricketts met Warder Clyde Allee, an ecologist who was undoubtedly the most important influence on him while at Chicago. Allee's intention, as he wrote, was to "investigate the relationships existing among the more loosely integrated collections of animals, which may rightly be designated as 'animal aggregations,' with regard to their ecological and behavioristic physiology, as well as with regard to their strictly social implications" (vii). His work began in 1911 and focused primarily on marine life in Woods Hole on Cape Cod, Massachusetts. There, Allee analyzed the grouping tendencies of various animals. In observing brittle stars, for instance, he noted their habit of intertwining their arms when gathered together under rocks. Allee eventually verified in his 1931 work, *Animal Aggregations,* that all animals, including man, tend to cooperate in nature, instinctually moving toward aggregation or a communal life. *Animal Aggregations* broke scientific ground by documenting aggregation patterns in "organisms whose interrelations have not reached the level of development usually called 'social,'" versus those noted in "organized societies, particularly those of mammals, birds, and insects" (3). Allee's work also set a precedent in ecological research: observations and data were gathered systematically on a yearly basis until patterns in animal behaviors and groupings were discerned.

Allee thoroughly contextualized his work in the related scientific discourse and included a comprehensive bibliography in *Animal Aggregations*. Early in the book, Allee stresses that his "present summary, gained from pioneering in this relatively new field, must be regarded as tentative in many respects [. . .] a point of departure, rather than a gathering of conclusions" (5). Ricketts was profoundly affected by Allee's theories and later used them himself as models when studying Pacific coast marine life. For example, in *Between Pacific Tides*, Ricketts cites Allee's study of brittle stars found in the Atlantic Ocean, comparing his teacher's findings to his own in the Monterey Bay. Allee had an impact on a number of students, including Albert E. Galigher, who, with Ricketts, had begun a small biological supply business as a means of support while in school. The two shared an apartment with another student, J. Nelson Gowanloch, who later became the chief biologist for the Louisiana fisheries department (Hedgpeth, *Shores* 1: 5).

Allee's influence on Ricketts's thinking was not limited to science. He "provided a philosophical foundation and a direction for a man [Ricketts] given to speculation, analysis, and synthesis" (Benson 193). Ricketts also applied Allee's concepts on group behaviors to humans, noting the significance of relationships between individuals and their environments and also among themselves. Ricketts believed that "everything is inherently related to everything else [. . .] and that to understand nature means to discern the relationship of its constituent parts" (Astro, *Steinbeck and Ricketts* 29). His belief in the importance of individuals integrating personal and social experiences in order to achieve a holistic awareness of the world and universe stems directly from Allee's teachings. Astro notes that Allee's work was a foundation for Ricketts, who often "leaves Allee far behind" in attempting to express his own holistic worldview, which he does "in terms more mystical than scientific" (28). Significantly, Ricketts's own theories on groups later influenced Steinbeck's fiction of the 1930s.

The early 1920s were marked by significant changes for Ricketts. He spent a total of three years taking classes at the University of Chicago without earning a degree; he was not concerned with graduating and never appeared to regret not having finished. In 1922 Ricketts met Anna Barbara Maker, who came to Chicago from Johnstown, Pennsylvania. After a short courtship, "Nan" and Ricketts were married on August 19, 1922. The couple moved in with the Ricketts family, and their son, Edward F. Ricketts Jr., was born on August 23, 1923. Soon after, Ricketts joined Galigher, who a few months earlier had moved to Pacific Grove, California. Together the two established Pacific Biological Laboratories on Fountain Avenue. Early in the century, the Monterey Bay was teeming with life, and the new partners had little trouble collecting specimens to sell to schools and laboratories across the country. Ricketts's fascination and diligence in observing and collecting was inexhaustible, and he spent hours on his own and with others scouring the shoreline. In her "Recollections," Nan Ricketts notes the wary reception that Ricketts, Galigher, and their business received from local inhabitants of Pacific Grove:

> The first year or so, the townspeople were very puzzled about what Pacific Biological Laboratories were, since we were out all hours of the night driving through the town in a noisy old Mitchell. Someone reported this to the police, because they were sure that the men were bootleggers, what with all the bottles and gallon jars we had. A policeman came to the Great Tidepool where Ed and Albert were collecting after midnight, and with flashlights at that, and asked them what they were doing. They said they were collecting specimens, which meant nothing to the policeman. Ed and Albert explained that they would preserve the specimens and send them to schools and universities for studies. He was satisfied that they were just some very odd people, and not bootleggers. But he did ask them to have the car repaired so it did not make so much noise going through town at night. It seemed that the muffler could be tended to. (23)

The lab and its owners were soon accepted in Pacific Grove, however, and over the next two decades Ricketts himself was recognized as an authority on the marine life of the region.

In 1925 or 1926 Ricketts's mother and sister moved to the area, followed by his father in 1927, and all lived in Pacific Grove. Frances worked at the lab and later moved to Berkeley, where she was employed at University Apparatus. Eventually she returned to Carmel to marry Fred Strong in 1931. For many years, Abbott Ricketts worked alongside his son in the lab assisting with the preparation of specimens for sale and shipment, which, as Nan wrote in her memoirs, "made Father Ricketts very happy" (51). Ricketts's partnership with Galigher had ended in 1924, leaving him as sole proprietor of the lab, and the Ricketts family often helped collect specimens to fill orders for schools. Throughout the 1920s, Ed worked hard to keep Pacific Biological Laboratories financially viable. In "Recollections," Nan notes that Ed's busy schedule "was never regular" but provided the family with a steady, if modest, income (5).

In addition to collecting for his business, Ed focused his attention on other projects. In 1925 he published two works: a short account of his cross-country walking tour of 1921 in *Travel* magazine and the first catalog for Pacific Biological Laboratories, which primarily targeted high schools and colleges, providing specimens for study and dissection. Significantly, the catalog reflects Ricketts's concern with conservation. He cautions: "It should be borne in mind (and this applies especially to local marine forms) that we must, above all else, avoid depleting the region by over-collecting. One or more formerly rich regions, according to reliable authorities, already afford instances of the ease with which depletion is brought about" (September 1, 1925).

Over the following two decades, Ricketts continued to study both the depletion and the abundance of marine life in Monterey Bay. Ed's intense concern for such ecological problems was well known in the community, and he was often sought out by students and researchers working at Hopkins Marine Station and by individuals interested in local marine life. Beginning in the 1920s, Ricketts studied plankton levels in Monterey Bay; he experimented with distilling sharks' liver oil for its medicinal value; he recorded the effects of wave shock and other physical factors that he found altered the ecological environment. In short, Ed was always a biologist, ecologist, writer, and thinker—it was this combination of talents that attracted people to him.

1930–1936: Ricketts, Steinbeck, and the Lab Group

In 1930 John Steinbeck and his first wife, Carol Henning, moved into the Steinbeck family cottage on Eleventh Street in Pacific Grove.

Frances Strong writes about the author's first encounter with Ricketts: "Ed and John did actually meet at the dentist. [. . .] Ed told me about his meeting with John in Dr. Curry's office and I remember all this vividly because I could not understand, when Ed and John had actually had had a real meeting which seemed to them significant, in the dentist's office, why John had to burlesque it later" (Strong to Hedgpeth, April 21, 1973).[1]

Biologist and critic Joel Hedgpeth, Steinbeck biographer Jackson Benson, and most other scholars assert that the now-infamous dentist office meeting was fictitious. They suggest instead that the men likely met for the first time at the home of Jack Calvin. Whatever their first point of contact, the men shared a complex friendship founded on mutual respect. Steinbeck and Carol spent much time at the lab, where they participated in discussions and parties with other local artists, scholars, and Bohemians drawn to Ricketts and the lab. In the early 1930s, Carol worked as Ricketts's assistant for a short period and often recounted to her husband interesting anecdotes about events and people she encountered at the lab (Hedgpeth, *Shores* 1: 12). It was not unusual for a casual evening gathering at Pacific Biological Laboratories to become a rowdy party, one or two even lasting days (Steinbeck later included these memorable parties in *Cannery Row* and *Sweet Thursday*). Steinbeck's experiences in Monterey in the early 1930s were some of the most profound in his life.

The friendship Ricketts and Steinbeck shared is of vital importance to any analysis of the men and their subsequent work. Most accounts of their relationship reveal the great extent to which Ricketts's own ideas and interests influenced those of his friend, particularly regarding science. Steinbeck's love of and fascination with marine biology and ecology, however, are widely known to predate his introduction to Ricketts, likely beginning with the class he enrolled in at Hopkins Marine Station in 1923. The class was taught by C. V. Taylor, who was largely influenced by William Emerson Ritter and his notions of "organismal conception of life" and the "superorganism" (Astro, Introduction xi). Ritter believed that when individuals form a group, the group essentially becomes a separate entity from those individuals of which it is composed: "wholes are so related to their parts that not only does the existence of the whole depend upon the orderly cooperation and interdependence of its parts, but the whole exercises a measure of determinative control over its parts" (Ritter 307). While Ritter's theory was meant to be applied to all organisms in nature, Steinbeck eventually became interested in its application to humans and how they interact among themselves. In the early 1930s, Steinbeck developed his "phalanx theory," which explored the dynamic between the group and the individual. He became fascinated by how "group man's" behavior differed from normal behaviors of the group's individuals. "Much of Steinbeck's thinking about the group man," notes Astro, "grew from his interest in the aggregational patterns of life in the tide pools," which he studied with Ricketts (*Steinbeck and Ricketts* 67).

Steinbeck developed his phalanx theory while nursing his dying mother in their family home in Salinas. One of his clearest statements about it is found in a letter to friend George Albee in 1933:

> [An] arrangement of cells and a very complex one may make a unit which we call a man. That has been our final unit. But there have been mysterious things which could not be explained if man is the final unit. He also arranges himself into larger units, which I have called the phalanx. The phalanx has its own memory— memory of the great tides when the moon was close, memory of starvations when the food of the world was exhausted. [...] And the phalanx has emotions of which the unit man is incapable. Emotions of destruction, of war, of migration, of hatred, of fear. These things have been touched on often.
>
> Religion is a phalanx emotion and this was so clearly understood by the church fathers that they said the holy ghost would come when *two or three were gathered together.* You have heard about the trickiness of the MOB. Mob is simply a phalanx, but if you try to judge a mob nature by the nature of its men units, you will fail as surely as if you tried to understand a man by studying one of his cells. (Steinbeck, *Life in Letters* 79–80)

Mobs are one type of phalanx, which Steinbeck explains indirectly in his short story "The Vigilante," which deals with a man's feelings after detaching himself from a mob. Steinbeck's interest in group man continued throughout the 1930s, and his novels *Tortilla Flat* (1935), *The Grapes of Wrath* (1939), and *The Moon Is Down* (1942) clearly show evidence of his abiding interest in the phalanx theory. When Danny and his friends are together in *Tortilla Flat,* they "became one thing [...] a unit of which the parts are men" (1), and individuals' characteristics are muted as the group's own personality emerges. In "The Leader of the People" in *The Long Valley* (1938), Jody's grandfather describes "westering" as the movement of "a whole bunch of people made into one big crawling beast" (224), an image that later appears in *The Grapes of Wrath* in Steinbeck's depiction of the migrants moving westward on Route 66. In chapter 14 of the novel, his description of individuals merging into group man explicitly incorporates Ritter's theories. "One man, one family driven from the land; this rusty car creaking along the highway to the west. I lost my land, a single tractor took my land. I am alone and I am bewildered. And in the night one family camps in a ditch and another family pulls in and the tents come out. The two men squat on their hams and the women and children listen. Here is the node, [. . .] this is the zygote. For here 'I lost my land' is changed; a cell is split and from its splitting grows the thing you hate—'We lost *our* land.' [. . .] This is the beginning—from 'I' to 'we' " (206). The movement from "I" to "we" echoes throughout the novel: from Casy inspiring Tom's interest in the labor movement to Ma guiding Rose of Sharon out of adolescent self-centeredness into the role of a selfless woman who eventually nurses a dying man for the greater good of "the people." Similarly, in *The Moon Is Down*, the townspeople act collectively to subvert enemy control, an idea Dr. Winter articulates: " 'The flies have conquered the flypaper' " (112). In these works, Steinbeck explores the phalanx in diverse situations, clearly showing that a group's personality and essence differ from the personalities of the individuals who compose the group.

Another significant idea Steinbeck undoubtedly shared with Ricketts was his concern with nonteleological thinking (also called "is thinking"). Ricketts defines the term as "concerning itself primarily not with what should be, or might be, but rather what actually 'is,' attempting at most to answer the questions what or how, instead of why—a task in itself rigorously difficult" (Hedgpeth, *Shores* 2: 162). Steinbeck was clearly influenced by this concept and integrated it into his fiction through characters such as Jim Casy in *The Grapes of Wrath* and Dr. Winter in *The Moon Is Down*. These characters do not seek causes for the present circumstances but an understanding of the true situation as it stands. Evidence of Steinbeck's interest in nonteleological thinking and the phalanx theory appears throughout his fiction of the 1930s.

Steinbeck and Ricketts also had an intense interest in holism, in integrating the human and the environment, and in seeing that a sense of the whole would lead to metaphysical speculation. Their fascination with metaphysics was undoubtedly ignited when Joseph Campbell came to Monterey in 1932, and for a brief period the three men shared an intense few months of conversation and speculation. Campbell moved into the house called Canary Cottage next door to the Ricketts in February 1932. He was soon befriended by Ricketts and became involved in socializing at the lab and in conversations with Steinbeck and Ricketts about group man and nonteleological thinking—particularly in relation to Steinbeck's work on *To a God Unknown* (1933). The three men had many common interests, including the Arthurian romance and mythology, and for a short period in the early spring of 1932, Steinbeck, Campbell, and Ricketts often engaged in discussions on transcendence and how, through momentary glimpses of truths beyond the temporal world, individuals live more fully, both emotionally and intellectually. Campbell was fascinated with the ideas Ricketts and Steinbeck shared with him about the phalanx and nonteleological thinking. These ideas, in turn, influenced his own concept of "transparent to the transcendent," or attaining transcendence which may be "glimpse[d] [...] ever and again (transparently, as it were) among the ordinarily opaque realities of our daily lives" (Larsen 258). Steinbeck found Campbell's insight constructive and seemed to regard his opinions highly; in fact, Steinbeck began revising his manuscript of *To a God Unknown* soon after meeting Campbell and listening to his comments about the novel's lack of sensuality and a "visual quality" (DeMott, Introduction xxviii). In his introduction to the novel, Robert DeMott notes that "Campbell discovered his and Steinbeck's mutual attraction to the 'tendency' of 'symbolism' as an essential way of apprehending and understanding reality" (xxviii). The three men spent a great deal of time discussing myth and symbolism in relation to *To a God Unknown,* in what became a period of personal and professional growth for all three.

Yet there was friction between Campbell and Steinbeck. Campbell and Carol, Steinbeck's wife, had engaged in a flirtation that culminated in an alleged confrontation between the two men, an incident Campbell recorded in his journal:

> I could feel the tension. I asked him to come in. "I'd like to talk a little, John," I said; "I was just going to see you."

John came in and sat heavily down, exactly where he had sat so rigid Friday night. He looked shot, and completely at a loss. . . .

"It's positively ridiculous even to think of my marrying Carol. The only question is, John, how I'm to withdraw from this mess with the least pain for her."

"And what about yourself?" asked John.

"To hell with myself," I answered. "I guess I ought to be able to tell my emotions what to do."

It pleased me to see that after that John was able to look me in the eye again. And I could feel that the sense of a rivalry had been broken into the warmer sense of a conspiracy against Venus. (qtd. in Larsen 193)

Although this journal entry suggests the men resolved the issue, they were never close friends. Despite their personal differences, however, they held a mutual artistic and professional respect for each other. In 1932 Steinbeck and Carol moved to Montrose, California, for seven months, returning after Campbell left for Connecticut. DeMott suggests that "Campbell's involvement with Carol was [. . .] a transactional event that injected a heightened sense of life into [Steinbeck's] novel" (Introduction xxix), asserting the positive result of this conflict on Steinbeck's work. Ricketts and Campbell's relationship did not seem to suffer despite the tension between Campbell and Steinbeck, and Ricketts and Campbell continued to spend a great deal of time together.

Campbell was also involved, albeit briefly, in Ricketts's passionate interest in marine biology, taking a trip with him in 1932. During that summer, Ricketts's personal life became turbulent when his wife, Nan, left Pacific Grove with their three children. While the couple never legally divorced, this became the first in a series of separations over the next few years that culminated in Nan moving permanently away from Pacific Grove in 1936 and finally beginning divorce proceedings in June 1946. Undoubtedly wanting to distance himself from such troubling personal problems and hoping to begin gathering material for a future project based on the coasts of Canada and Alaska, Ricketts decided to go on a collecting expedition to Juneau, Alaska, with Campbell and a group of friends.

In June 1932, Jack Calvin, his wife, Sasha, Campbell, and Ricketts traveled to Seattle, where they set out on Calvin's thirty-three-foot boat, the *Grampus*, to Sitka, Alaska, where Sasha's sister Xenia Kashevaroff joined the voyage. In addition to collecting, the group discussed books, philosophy, and ecology. "The conversations or bull sessions," writes Hedgpeth, "often lasting until three AM, were about Spengler, science as a way of life, religion, men, women, people, marine biology, almost everything; and when they were not talking, they were reading" (*Shores* 1: 14). In an unpublished typed fragment titled "Worth Remembering Concepts Arrived at or Considered during the 1932 BC Alaska Trip," Ricketts wrote:

> Instinct in animals is possibly a pattern arising from the sensations etched on an unconscious memory. And triggered when needed by the associations (also probably unconscious) arising from the repetition of a particular sensation. Available at need (rather than at will) because not subject to the blinding light of intellectualism. The old figure of speech of the stars representing the emotion or instinct pattern which is made invisible by the sunlight (intellect) altho still present in the background. A folk geist thus formed could be instinct. The unconscious and "short cut" feeling release of this wisdom by the emotional modifications of the individual equals intuition. The conscious and rational tapping of this equals intellect.

Ricketts's fascination with instinct and intuition derive in large part from his extensive study of Carl Jung's theories, particularly *Contributions to Analytical Psychology* (1928). At about this time, Ricketts befriended Evelyn Ott, a psychiatrist from Monterey who had studied with Jung, and he talked at length with her about Jung's theories and dreams. Campbell, too, studied Jung, and he and Ricketts likely shared intense discussions, both on the *Grampus* and in Monterey, about ideas such as unconscious emotion, or instinct.

Significantly, Allee also considers the concept of instinct in *Animal Aggregations,* although not deeply. He feels it is a difficult aspect of animal behavior to study: "[W]e shall avoid dwelling on the aspects of behavior usually called 'instinctive,' [. . .] not due to a disbelief in the reality of instinctive social behavior, but rather to a conviction that progress lies in a field where the elements of behavior can be more exactly ascertained" (11). Clearly Allee's theories were never far from

Biographical Essay / xxix

Ricketts's mind, and these undoubtedly were also part of the conversations Ricketts shared with Campbell, Ott, and others.

The complete log Ricketts kept on the Alaska trip was prepared into a typescript upon his return to Pacific Grove. He titled it "Notes and Observations, Mostly Ecological, Resulting from Northern Pacific Collecting Trips Chiefly in Southeastern Alaska, with Special Reference to Wave Shock as a Factor in Littoral Ecology." While the typescript includes much scientific data, it also reveals Ricketts's appreciation for the beauty of the landscape. He writes, "The attenuated dawns and twilights, the continued drizzly rain, and the thrushes singing for hours at night and morning from the wet and steep hillsides—the only sound in this quiet region, aside from the rush of waterfalls—are the things I remember chiefly from this country" (qtd. in Hedgpeth, *Shores* 1: 16). Hedgpeth notes that "Ricketts later planned to salvage [selections from the typescript] for his projected book on *The Outer Shores*" (2: 15). He never completed the manuscript, however, which was to have been a study of the littoral of the northwest Pacific coast.[2]

When the group returned to California, Ricketts and Campbell spent the rest of the summer socializing and collecting locally. Campbell eagerly joined the lively circle of friends surrounding Ricketts, although he was more often an observer than a participant in large gatherings. At times, though, he was swept up in the fun, once even becoming "engaged in a face-slapping contest with Rich [Lovejoy] on the floor." Campbell wrote in his journal: "When we had tired of that I delivered a speech in German. Tal [Lovejoy] got down on the floor and we pretended to gaze each other's hearts out" (Larsen 181–82). Although Campbell lived in Pacific Grove for only a few months, leaving in the summer to begin teaching at the Canterbury School in New Milford, Connecticut, he and Ricketts formed a deep connection that they maintained through their letters from 1939 until Ricketts died nine years later.

Throughout the early 1930s Ricketts explored and studied the Pacific coast, taking trips from Ensenada, Mexico, to Juneau, Alaska. Monterey Bay, however, was always his primary focus. Friends "who enjoyed going with him on low tide collecting trips, [and] encouraged him to write all these things up," helped inspire him to begin a book concerning primarily Monterey Bay littoral (Hedgpeth, "Philosophy"

104). Thus, throughout the late 1920s and early 1930s he began assembling material for *Between Pacific Tides* (1939), a handbook of marine life along the northern Pacific coastline. Its vision was—and remains—innovative. The book was arranged according to ecological habitat instead of species. For example, the first section examines animals found in the "Protected Outer Coast," including "Rocky Shores" and "Sandy Beaches" (Ricketts, *Tides* xi). Significantly, Ricketts was familiar with an earlier study on which he may have modeled the structure of his book. In Verrill and Smith's 1872 "Report on the Invertebrate Animals of Vineyard Sound and Adjacent Waters," they, too, organized their study of animals according to habitat. Ricketts included the report in the bibliography to *Between Pacific Tides*, and as it surveys invertebrates found primarily in the Martha's Vineyard region off the coast of Massachusetts, his only interest in it was likely its organization. Ricketts's concern with organization is significant. He wanted the book to be accessible to both scientists and the general public.

Producing *Between Pacific Tides* was, in many ways, a collaborative effort. Friend Jack Calvin assisted Ricketts in collecting, proofreading, editing, and providing photographs, while local artist Ritch Lovejoy contributed detailed drawings of animals. Edward Jr. helped compile data for and create many of the book's graphs and tables. He remembers the painstaking hours he spent making a circular-shaped chart showing the seasonal diatom production in the Aleutian Islands and southern California and how his father always had to double-check each calculation himself, not wanting to risk a mathematical error in the book (275). Ricketts's concern with ecological holism, which he considered the interrelation of animals to each other and their environment, which can "lead us to the borderline of the metaphysical," set the book—and himself—apart from the norm (60). In fact, Ricketts's visions not only of the book but also of scientific methods in general were quite untraditional, as seen in this excerpt from some of his early notes. "The whole picture should be stressed, and one's feet should be kept on the ground by frequent actual collecting, by observing how the animals live and by constant open-minded checking. Because too often (in zoology as in all other fields) what are thought of as disciplines operate chiefly as biases—prescribed ways of thinking and of doing, into which the professional may retreat when shocked or challenged by some anomaly" ("EFR Essay No. 2"). The above passage may be considered Ricketts's own philosophy not only of science but also life as a whole as he avidly explored and collected along the Pacific coast and remained open to innovations and variations in his work as a scientist. Unfortunately, Ricketts's life was about to be challenged in a most profound manner.

1936–1939: Ricketts's Philosophical Essays

In 1936, after a permanent separation from Nan, Ricketts moved into the lab at 800 Ocean View Avenue, working and living comfortably for only a few months. On November 25, 1936, an electrical fire began in the Del Mar Cannery next door to the lab. It destroyed buildings on Cannery Row, including the lab and all of its contents. Given the incredible volume of writings—essays, notes, and correspondence—that exist from the twelve years following the fire, estimating the loss of prefire material is next to impossible. Edward F. Ricketts Jr. confirms that his father kept carbon copies of virtually all of his correspondence and essays in the lab, and estimates that hundreds of letters must have been lost. "It would have been an incredible amount," he said (Interview June 28, 2000).

Perhaps the most devastating loss of property in Ricketts's estimation, however, was his library. He had compiled an extensive collection of scientific books, journals, and articles throughout the years, as well as numerous volumes of poetry and other literary texts, which he spent much of the rest of his life reassembling. The only property Ricketts managed to save included his car, a portrait of himself by friend James Fitzgerald, and his typewriter. The financial loss was estimated to be close to twelve thousand dollars; however, Ricketts collected only three thousand dollars in insurance money. Although a lawsuit was later filed with Pacific Gas and Electric, no additional compensation was ever paid to Ricketts. Necessarily, his attention turned toward rebuilding the lab, his home, and his workplace.

The new building was much the same as the old, with one notable addition being the bank of windows looking out onto Ocean View Avenue. Friends contributed time and ideas to the construction of the new lab and threw rehabilitation parties for Ricketts, giving him books in an attempt to help rebuild his lost library. The first and perhaps most notable of these book showers was at Christmastime in 1936, just one month after the lab was destroyed. His correspondence from that period reflects his gratitude. Consequently, the existing volume of letters from after the fire begins with numerous references to the losses Ricketts suffered and the rebuilding that ensued.

In addition to his correspondence, much of the research material Ricketts compiled for *Between Pacific Tides* was destroyed. Thankfully, the book had been accepted by the Stanford University Press just months before the fire, and the typescript and plates were already with the publisher. For years, Ricketts had encountered a great deal of resistance from Stanford University Press, who was wary of his lack of a university degree. Despite some initial reservations, the press did, in fact, accept the book, although they were disappointingly slow to publish it. Several of Ricketts's earliest letters in this collection record his frustration at the various delays in publication. In a letter to friend and Swedish marine biologist Torsten Gislen, Ricketts writes: "Nothing recent from Stanford Press about the book. It has been scheduled for issuance a dozen times during the past few years, but they are slow. Our contract reads 'Within a reasonable time'" (March 31, 1937). *Between Pacific Tides* was eventually published in 1939, receiving acclaim in the scientific community. Albert Campbell's review in the *Monterey Peninsula Herald* quotes Dr. W. K. Fisher, director of Hopkins Marine Station, as saying, "we intend to have it in our library here at the Marine Station and to use it for instruction." *Between Pacific Tides* is in print today and continues to be used in college classrooms. Ricketts remains recognized as an innovative scientist whose thoughts, as oceanographer James C. Kelley notes, "contain all of the primary elements of what 'New Age' writers, thinking they have found something new and revolutionary, call 'deep ecology'" (28).

Ricketts likely worked as diligently on a series of three philosophical essays as he did on his exploration of Pacific littoral zones. Throughout the 1930s he repeatedly revised these three significant essays that contain much of his philosophical thought. Although no exact date of their origin is known, various drafts circulated among friends throughout the decade, and in 1939 Ricketts attempted to publish them. Steinbeck helped him, using his status to find a publisher. Ricketts wrote to Campbell on October 7, 1939: "I have started attempts toward publishing those three essays. Sent them first to *Harper's* and they came back just a couple of days ago, and only a few days after Jon, who was with de Kruif in Michigan, had wired me for dope on the one on non-teleology and had then, with de Kruif, unbeknownst to me, wired to *Harper's* asking them to make a typed transcription of the essay charged and sent to Jon. He apparently thot that that influence would call the things to their attention, in addition to his wanting a copy of the revised essay." Unfortunately, his essays were not published, in spite of Steinbeck's efforts.

In the first essay, Ricketts defines the notion of nonteleological thinking, or "is thinking," which is achieved when one does not seek causes or reasons. Ricketts poses various questions—such as "Why are some men taller than others?"—and subsequently demonstrates the processes of answering them first teleologically and then nonteleologically. The teleological answer to the above question, for instance, would be "the under-functioning of the growth-regulating ductless glands," while the nonteleological response would be that "there can be no 'answer.' [. . .] There can be only pictures which become larger and more significant as one's horizon increases" (Hedgpeth, *Shores* 2: 163).[3] Nonteleological thinking was likely an amalgamation of Ricketts's interests in Jung, mythology, ecology, and philosophy. He constantly revised and refined his ideas based on discourse with friends through letters and discussion, attempting to better articulate his thoughts. Ultimately, Ricketts believes, nonteleological thinking enables an individual to live more fully: as questioning decreases, acceptance increases, and a person moves closer to holistic awareness, or transcendence.

Ricketts found evidence of nonteleological thinking virtually everywhere. "The deep underlying pattern inferred by non-teleological thinking crops up everywhere—a relational thing surely, relating opposing factors on different levels, as reality and potential are related. But it may not be considered as causative, it just exists, it *is*, things are merely expressions of it. [. . .] The whole picture is portrayed by *is*, the deepest word of deep ultimate reality, not shallow or partial as reasons are, but deeper and participating, possibly encompassing the oriental concept of 'being'" (170). Ricketts's concern with participating is derived from the Taoist concept of "quietism," in which surrendering oneself to the temporal world's limitations brings enlightenment. Ricketts studied Taoism, particularly the 1935 edition of the *Tao-te Ching,* a gift from a Christmas book shower in 1936, and incorporated a number of tenets into his personal philosophy. Significantly, Joseph Campbell had been studying Taoism during his stay in Monterey in 1932, which no doubt would have been a topic of discussion with his friend. Ricketts believed that through quietism and nonteleological thinking, an individual may become less concerned with causes and participate more fully with the universe, hence achieving a sense of "holism," which he defines in a letter to Campbell as "a light; the integration of the parts being other (and more than) the sum of the parts; an emergent" (July 3, 1940). Two years later, Ricketts described holism to friend Heinz Berggruen, to whom he noted: "thinking, feeling, intuition and sensation all contribute, and in which they all merge" (July 11, 1942). Clearly, Ricketts's ideas about nonteleological thinking also influenced Steinbeck, particularly in *Of Mice and Men* (1937), which Benson noted was an "attempt to write from a completely non-teleological standpoint: no cause and effect, no problem solution, no heroes or villains" (327). He originally titled the novel "Something That Happened."

Ricketts believed one may find evidence of nonteleological thinking and holism in the work of various artists, particularly poets, and he explored these connections in a second essay, "The Spiritual Morphology of Poetry." In it he defines four categories, or "growth stages," into which all poets may classified. The first consists of the "naïve poets," "and in its most primitive phase extols pastoral beauty. Their poetry involves a simple and fresh statement of the joy of existence. [...] Most western poetry falls in this group" (Hedgpeth, *Shores* 2: 84). Ricketts considered Christopher Marlowe, William Shakespeare, Samuel Johnson, William Wordsworth, Percy Bysshe Shelley, and William Yeats, among others, to be naive poets. John Keats, Matthew Arnold, and Alfred Tennyson he considered to be in the second group: those who "achieve an *emergent* of thought or feeling quite different from the lessons which the less significant poets are always drawing from their subject matter" (84). Ricketts often referred to these as the "more sophisticated poets," and he believed they comprised a small group due to its "crucial and transitory character. Most of the truly great men who achieve the need for critical introspection either cannot pass the test, and so are destroyed by it without having created any lasting expression, or go through it to come out on the other side" (85).

The "mellow poets"—or those who have "come out on the other side"—"extol ugliness, tragedy, even the clay feet, not for themselves, but because they are vehicles of that beyond quality, the significance of which they have come to realize [. . .] although they may be unable to state" (87). Walt Whitman, Robinson Jeffers, Ezra Pound, and "most Chinese poetry" fall into this third category, according to Ricketts. In his estimation, these poets have come closest to expressing the moment of "breaking through," or transcendence, at which point integration of the visceral and the spiritual is achieved. "Black Marigolds," the anonymous Sanskrit poem read by Doc in Steinbeck's *Cannery Row* (1945), is also included in this category, which Ricketts considers "one of the greatest poetic expressions, even in translation, that I have ever encountered" (Hedgpeth, *Shores* 2: 87). The final classification, or the "all-vehicle mellow poet, [has] not yet emerged at least in this culture" (87). Such a poet would, in Ricketts's theory, speak out of "the heaven glimpsed by his predecessors, the heaven-beyond-the-world-beyond-the-garden, [. . .] would know 'it's right, it's alright': the 'good,' the 'bad,' whatever *is*," and break through (87–88). This point may only be reached through nonteleological thinking or quietism, which would allow the poet's mind to stop reaching beyond life as it is, and, in turn, achieve transcendence.

Ricketts's third essay is perhaps the most significant, as it further explains his notion of transcendence, or enlightenment. In "The Philosophy of 'Breaking Through,'" he widens the scope of his consideration from poetry to include literature, art, music, and philosophy. He sees "breaking through" as an "un-named quality" that emerges through the words, paint, and/or music that act as vehicles of individual works. This quality may be perceived only in flashes, such as when Ricketts heard "Madame Butterfly" for the first time and "there was the same quiet realization that no sacrifice [. . .] would be great enough if one could only share that girl's sorrow" (70). In an instant of connection—or integration—between an individual and the universe, a flash of breaking through, or transcendence, may take place. "It's all part of one pattern," Ricketts writes. "I suspect now that the pattern is universal, that we fail to see the transcending simplicity of it only because of obstacles on our inward horizons" (71). One of the most significant explorations of the concept of breaking through is found in *Sea of Cortez* (1941), the result of Ricketts and Steinbeck's expedition to Baja California in 1940. In chapter 14 (Easter Sunday), through nonteleological thinking, the men approach the sort of transcendence Ricketts describes in his essay. "This little trip of ours was becoming a thing and a dual thing, with collecting and eating and sleeping merging with the thinking-speculating activity. Quality of sunlight, blueness and smoothness of water, boat engines, and ourselves were all parts of a larger whole and we could begin to feel its nature but not its size (151). Steinbeck and Ricketts spent time during their Cortez expedition and in Monterey discussing Ricketts's philosophies and essays.

Ricketts continually sought evidence of breaking through in the world around him, finding it at times in Steinbeck's work—he mentions *To a God Unknown* repeatedly in "The Philosophy of 'Breaking Through'" as a literary example—but also in music and visual art. Monteverdi's madrigals, Bach's *Art of the Fugue,* and Gregorian chants affected Ricketts profoundly. Most people who spent time at the lab vividly remember music being played. Ricketts listened to music as attentively as he studied the tide pools. Michelangelo, Leonardo da Vinci, and other Renaissance painters "tower as giants clear to this day," Ricketts noted in his essay (Hedgpeth, *Shores* 2: 76). He periodically ordered from museums prints of paintings by artists such as Paul Cézanne, Pablo Picasso, and El Greco. In a letter to friend Xenia Cage, wife of experimental percussion composer John Cage, Ricketts comments on Russian figurative painter Aleksei Yavlensky: "I can't get that one head (the pastel colored one) out of my head; or rather I can't get the idea out, or the idea of the idea, something as involved as that, since I've lost memory of the appearance of [the] picture itself" (October 9, 1939).

In addition to referencing influential artists, Ricketts's letters from the last twelve years of his life also trace the various revisions his essays underwent and his eagerness to see them published. Campbell read the essays, as did Henry Miller after his visit in 1941; friends and family also read the essays, and Ricketts's process of rewriting continued throughout the 1940s. He hoped to integrate the three essays into a single volume. In a notebook he began in 1940, he outlined its structure:

Structure of the series of essays called "Participation":
1. Being. "Non tele-thinking" (I had thot of phil of br. thru as 1st previously. In a way it makes no diff. tho)
2. Becoming. "Phil of Breaking Thru"
3. The working out of these two things in an actual, practical way in life [that is, in anything in life; Literature, science, human relation—anything] I chose poetry. ("New Series, 1" 9)

With the exception of "Non-teleological Thinking," which was included in *Sea of Cortez*, Ricketts's philosophical essays were not published in his lifetime. Significantly, Ricketts's notebooks combine personal, professional, and philosophical ideas and experiences. Even his most formally written scientific documents include both direct and indirect references to these three essays. Just as he sought to integrate all experiences in his life, Ricketts's writing merged the ideas that most fascinated him.

1940: The Sea of Cortez Trip

A few months before the lab fire of 1936, John and Carol Steinbeck had left Pacific Grove and moved to Los Gatos, California; in doing so, the author left Ricketts's immediate and intimate influence. Steinbeck's next project was "The Harvest Gypsies," a series of articles for the *San Francisco News* about migrant workers in Central Valley. The series helped prepare him for writing *The Grapes of Wrath* (1939). Work on the novel was intense, and Steinbeck was physically and emotionally drained upon its completion. "Now I am battered with uncertainties," he wrote on October 16, 1939, in his journal.

> That part of my life that made the *Grapes* is over. I have one little job to do for the government, and then I can be born again. Must be. I have to go to new sources and find new roots. I have written simply for simple stories, but now the conception and the execution become difficult and not simple. And I don't know. I don't quite know what the conception is. But I know it will be found in the tide pools and on a microscope slide rather than in men. (DeMott, *Working Days* 106)

/ *Biographical Essay*

Steinbeck soon connected with Ricketts and embarked upon an ambitious project, a new conception conceived in three phases. The first was a handbook of marine life in San Francisco Bay; the second, an exploration of the littoral of the Sea of Cortez; and the third, a study of the Pacific northwest coast, which Ricketts referred to as "the outer shores." Thus, Ricketts would have mapped the whole of the Pacific coast with Steinbeck's help. Although never fully executed, the book Steinbeck and Ricketts did in fact write proved to be Ricketts's most significant contribution to literature: *Sea of Cortez*.

Plans began for the handbook of San Francisco Bay in late 1939. In a letter written to Campbell in October 1939, Ricketts wrote, "I am to do another job for Stanford—a manual of the invertebrates of the SF Bay area. Won't take long, a year or so if I can get to it" (October 7, 1939). The project was to instruct Steinbeck in the ways of scientific study, and, as he wrote to friend Carlton Sheffield in December 1939, "build some trust in the minds of biologists" (Steinbeck, *Life in Letters* 196). Steinbeck studied and read under Ed's direction, and they began making short trips to the bay, surveying the area and collecting in preparation for the handbook. Steinbeck bought and outfitted a truck for the impending expedition, much in the way he would "Rocinante" years later for his trip that inspired *Travels with Charley*, published in 1962 (Benson 428).

The men worked diligently throughout December 1939 and January of the following year, each writing sections of text that would eventually be combined. The "Bay Area Handbook" was to be for high school and college students and would include a catalog of animals, explanatory essays, and an annotated bibliography. Ricketts drafted a "Suggested Outline for Hndbk of Marine Invert. of SF Bay Area": "The object of the book is to provide handy, easily consulted, collecting handbook, to be carried during collecting or observation for identification of the commoner forms of marine invertebrates. It will be specifically designed for beginning biology classes but will be written and ordered so that it may be used by the sea coast wanderer who finds interest in the little bugs and would like to know what they are and how they live. Its treatment will revolt against the theory that only the dull is accurate and only the tiresome, valuable." At times, Steinbeck struggled with his portion, writing in his journal in January 1940: "Came down here [to Pacific Biological Laboratories] to try to work on the tide pool hand book. I discover that there are no easy books to write and that this may well be one of the hardest" (DeMott, *Working Days* 109–10). In fact, Steinbeck completed at least two drafts of what would have been the preface to the book, one of which was later included as the foreword to the revised edition of *Between Pacific Tides* (Hedgpeth, *Shores* 1: 36). Steinbeck was, however, excited about the project and eager to begin exploring, collecting, and preserving. Edward F. Ricketts Jr. remembers: "He was like a kid. Once he and I were trying to figure out how to preserve anemones that Dad kept in tanks in the lab. We got in his car and went to buy dry ice thinking we could quick freeze them. We dropped the ice into the tanks, not knowing what to expect. [. . .] Then Dad came in wondering what was going on; the carbon dioxide from the dry ice was choking him" (Interview May 11, 2001).

Ricketts's notebooks contain lists of books to order, notes on necessary supplies, sketches of equipment to be bought or—as with a portable saltwater pump and refrigeration system—to be built by himself. These notebooks also contain personal entries, including accounts of vivid dreams, thoughts on his philosophy of breaking through, and conversations with Steinbeck regarding their plans. Many entries are reflections about a married woman with whom Ricketts shared a passionate, if turbulent, love affair. Despite the fact that Ricketts rarely mentioned her in letters to friends and family, his notebooks and papers reveal the intensity of his feelings and his struggle to move on after she ended their affair. Clearly, he channeled the energy from his emotions into planning for the handbook. When Ricketts and Steinbeck began their collaboration, their attention was focused on the future, and they were both eager to begin research for what each hoped was a second successful text.

Steinbeck and Ricketts also began planning a study of the littoral of Baja California, Mexico. In March 1940, the Bay Area expedition was abandoned in favor of the Mexico trip. Hedgpeth notes that in late 1939 Ricketts had planned a trip to southern California for the spring of 1940, which likely "grew into a full-scale collecting trip or 'expedition' by boat" (*Shores* 1: 32, 41). As their attention turned from San Francisco Bay to Mexico, Ricketts and Steinbeck began to write letters to Mexican and U.S. officials, hire and outfit a boat, and locate a crew for the expedition. Finally, the *Western Flyer*, a seventy-six-foot

xl / *Biographical Essay*

fishing boat, was hired along with its captain, Tony Berry, and a crew consisting of engineer Tex Travis, Sparky Enea, and Tiny Colletto. Carol Steinbeck went along as the cook, although she and Steinbeck were having marital difficulties; in fact, the trip was looked at as an effort to save their failing marriage. With a crew of seven, the *Western Flyer* left Monterey on March 11, 1940, and the group spent about six weeks exploring, observing, and collecting in the Sea of Cortez.

While the expedition was clearly a collaborative endeavor, Ricketts and Steinbeck prepared for the trip in different ways. In *Parallel Expeditions: Charles Darwin and the Art of John Steinbeck* (1995), Brian Railsback notes the significant influence of Darwin's *Beagle* voyage (1831–36) on Steinbeck. The author was fascinated, Railsback observes, with "Darwin's methods and theories, as well as his contribution to holistic thought. [. . .] Of all the biologists and scientists that Steinbeck read, only Darwin inspired such personal affection" (36–37). In fact, in the first pages of *Sea of Cortez*, Steinbeck wrote, "Our curiosity was not limited, but was as wide and horizonless as that of Darwin or Agassiz or Linnaeus or Pliny" (2). Ricketts also prepared for the trip by reading extensively, and the books and authors he lists in his notebook reflect his broad scope of interests: Hemingway, Keats, and Willa Cather are listed among ecologists and biologists. The trip was not meant to be purely scientific; both Steinbeck and Ricketts saw it as a hands-on exploration of environment and philosophy—an opportunity to fully experience their surroundings to achieve a sense of holism.

The forty-one-day trip covered approximately four thousand miles of coastline. While the crew of the *Western Flyer* had some initial reservations about Ricketts and Steinbeck's plan, it was not long before everyone became actively involved in collecting and recording data. Steinbeck and Ricketts agreed before the trip to keep separate notebooks, with the understanding that upon their return Steinbeck would combine both into a collaborative text. Steinbeck himself did not maintain a journal, however, and subsequently pieced together the log portion of *Sea of Cortez* from a typescript of Ricketts's notes and Captain Berry's log. The fact that Steinbeck did not keep his own written record of the trip does not indicate a lack of enthusiasm. In their desire to "see everything [their] eyes would accommodate," the men did not limit themselves to gathering only scientific data, but experienced as many aspects of the environment as possible (2). Ricketts's notes from the trip reflect this and are a combination of scientific observations, including sketches and measurements, along with more personal and casual comments about the people he encountered. They also reveal his kindheartedness and sense of humor. For example, on March 22, he writes:

> La Paz. Sand to muddy sand to sandy mud at El Mogete, 1 mile N of La Paz. The important items were: Cerianthus, Dentalium, bit sipunculid and new cucumbers. When we returned, 40,000 kids were on the boat awaiting us with the following animals. [. . .] 4 things today: Church, good friday [*sic*], Viernas Santo, 11 to 12, women in black widow's clothing, smell of people and of perfume, some blondes, fine earnest priest black eyes blazing, good voice, old Spanish chants like madrigals with quarter tone wails, chorus of children's voice singing loud and out of tune on the notes they know, coming in pretty strongly on some of the obvious or tuneful phrases. ("Verbatim Transcription" 10)

Some entries, however, are more philosophical. On March 12, 1940, Ricketts writes, "The working out of the ideal into the real, [is] a constant process, and the relationship between inward and outward, or microcosm and macrocosm, are all here, as they are everywhere" ("Verbatim Transcription" 2). *Sea of Cortez* captured this combination of the scientific, personal, and philosophical to convey ecological and experiential holism.

In the first pages of the book, Steinbeck outlines their purpose. "We have a book to write about the gulf of California. We could do one of several things about its design. But we have decided to let it form itself: its boundaries a boat and a sea; its duration a six weeks' charter time; its subject everything we could see and think and even imagine; its limits—our own without reservation. [. . .] We wanted to see everything our eyes would accommodate, to think what we could, and, out of our seeing and thinking, to build some kind of structure in modeled imitation of the observed reality" (1–2). This consideration of the interconnectedness between man and his environment—or the microcosm and macrocosm—in *Sea of Cortez* is the most explicit written record of ideas prevalent throughout both men's writings, most notably in Ricketts's "Non-teleological Thinking" and "The Philosophy of

'Breaking Through'" and in Steinbeck's fiction, from *To a God Unknown* to *Cannery Row*.

Yet *Sea of Cortez* was not merely a log. After their trip, Ricketts set about the task of finishing the annotated phyletic catalog and bibliography for the book. The result was a more than three-hundred-page taxonomic list of specimens collected on the expedition; animals are listed by phylum, and an entry for each includes annotated references to sources of information about that animal. In the introduction to the catalog, Ricketts wrote: "Although our purpose was, primarily, to get an understanding of the region as a whole, and to achieve a totopicture of the animals in relation to it and to each other, rather than to amass a great collection of specimens, we nevertheless procured more than 550 different species in the pursuit of this objective, and almost 10% of these will prove to have been undescribed at the time of capture. And we merely scratched the surface" (304). With the discovery of at least thirty-five new species, the trip and the catalog are significant contributions to marine ecological studies of the Pacific. Both Ricketts and Steinbeck believed the book needed high-quality images of the specimens and fought Viking Press's reluctance to spend money on color plates. Ultimately, Viking agreed to include the images, and Ricketts and Steinbeck were extremely proud of the volume, which they felt well represented their experience in the Gulf.

The expedition brought immense satisfaction to both men: Ricketts had now conducted an in-depth ecological study of a second Pacific coast region (the first having been that covered in *Between Pacific Tides*), and Steinbeck had shifted his creative energies from fiction writing to scientific exploration. The trip was clearly of great significance to Steinbeck, who later wrote to Pascal Covici, "I think I will be looking back longingly on the Gulf of Lower California—that sea of mirages and timelessness. It is a very magical place" (*Life in Letters* 221). They did not, however, know that more than fifty years later their study of the Gulf of California would still be considered one of the most comprehensive accounts of the region.

1940–1942: Mexico City and Monterey

Upon their return from the Sea of Cortez, Steinbeck and Ricketts soon traveled to Mexico again, although, due to a crucial difference in their outlooks, they did not work collaboratively. The *Western Flyer* and her crew returned to Monterey in May 1940, and shortly thereafter Steinbeck and Carol flew to Mexico City to begin filming *The Forgotten Village* (1941) with Herb and Rosa Kline. Ricketts soon followed, bringing Steinbeck's car over the border and assisting with the production. He wrote to friends Ritch and Tal Lovejoy from Mexico: "I received a not very warm welcome—a possibility I had anticipated. [. . .] I have seen almost nothing of Jon or Crl. I am off their list. [. . .] A new experience for me being the poor cousin" (June 18, 1940). Ricketts met this new experience with acceptance and reflected in his journal on his ability to keep a positive perspective throughout such situations. "There are things in which I excel: My sense of truth, my ability to understand a situation or a thing or a person or group. Of these things, my ability to go to the depths and fit them all into a unified field; to collect and collate information and to get the whole picture out of it" ("Second 1940, Notebook One" 23). So Ricketts took advantage of the resources in the university library in Mexico City to expand the bibliography to be included in *Sea of Cortez* and spent his leisure time exploring the city. "I have a personal affection for this hot country, as one would have an affection for a girl" (20).

The coolness between Ricketts and Steinbeck was caused by the theme of *The Forgotten Village*. The film depicts the conflict within a native Mexican village when its children become ill due to contaminated water. Steinbeck's protagonist, young Juan Diego, seeks help from his teacher and doctors from the city after the traditional folk remedies of the village *curandera* (medicine woman) do not help his dying brother. Steinbeck advocated the benefits of bringing modern medicine to the village; Ricketts, however, vehemently disagreed with his friend, believing instead that the native culture must not be disturbed by outside influences. As a result, the two spent much of their time in Mexico apart from one another. Of central importance on the trip, however, was an essay Ricketts wrote while in Mexico, an essay that outlines the essential differences between Steinbeck's and his own views on nonteleological thinking. The resulting antiscript in response to *The Forgotten Village*, titled "Thesis and Materials for a Script on Mexico: Which Shall Be Motivated Oppositely to John's 'Forgotten Village'" is a defense of the native culture of the village, particularly the importance of the curandera. Ricketts argues for preservation of the traditional when imposed upon by the modern: "The gradual influence of the 'new thing,' which was at first outside the country, [...] is now working from within, in corrupting the old life and in upsetting the age-old relation between the people and the land and between the people themselves" (Hedgpeth, *Shores* 2: 179). For Ricketts, any progress that undermined a traditional culture was suspect. His antiscript keenly discerns the larger picture—or "toto-picture," a term he often used to describe a sort of Taoist holism—of what is happening in *The Forgotten Village*. Fortunately, the disagreement between Steinbeck and Ricketts did not create a permanent rift between them; instead, both accepted each other's positions and let the matter go. Ricketts wrote to Ritch and Tal Lovejoy that "being good natured and not holding grudges very much makes it not bad at all" (June 18, 1940).

After returning to Cannery Row, Ricketts immediately began to incorporate the bibliographic material from his research in Mexico City into *Sea of Cortez*. Steinbeck returned to Mexico in the fall of 1940 to finish filming *The Forgotten Village*. His start on the log portion of *Sea of Cortez*, therefore, had to wait until January 1941. After returning from Baja, Ricketts had begun a relationship with Toni Jackson, whom he met through a mutual friend, Virginia Scardigli. Toni, divorced and with a small daughter, Kay, eventually moved into the lab on June 10, 1940, and lived with Ricketts as his common-law wife until 1947. An intelligent, engaged woman, Toni worked as a writer and reporter for local newspapers and magazines and often traveled with Ricketts on his collecting trips. She later worked intermittently as a typist for Steinbeck and typed both the log and catalog of *Sea of Cortez*.

Toni was a spirited addition to the lab group, and she and Ricketts socialized with some fascinating visitors in the 1940s. An interesting meeting occurred in the summer of 1941 when writer Henry Miller visited Monterey with his friends Gilbert and Margaret Neiman. Miller was introduced to both Steinbeck and Ricketts and took an immediate liking to the latter. Of Steinbeck, Miller was not so fond, as evidenced in his letters to Ricketts. Miller was also impressed with Toni, and the two exchanged letters for a short period after Miller's departure. Although Miller did not visit Ricketts in person again, the two continued to correspond irregularly over the next few years. Their letters discuss Miller's writing and Ricketts's essays, along with general news about mutual friends, such as the Neimans. An interesting passage about

Miller, titled "The Position of Henry Miller and How It Came About," is found in one of Ricketts's notebooks from 1944–45. "He's hated or idolized—rarely fairly: and for only a small part of his work: the so-called pornographic aspects. [. . .] Sex was for him also a vehicle for 'breaking thru.' [. . .] I should say that his feelings are well developed, he is in one sense a minor saint. But not wonderfully developed or disciplined along intellectual lines, altho he had a fine mind to build with. So he's not nearly so good or so bad an artist as he's painted, a terribly good person, a fair but very honest artist with a [. . .] flair for imagery" ("New Series, 3" 86).

In addition to socializing with Miller and numerous others, Steinbeck and Ricketts made their final additions and corrections to *Sea of Cortez* throughout the fall and winter of 1941. Viking then released the book in the first week of December. The bombing of Pearl Harbor on December 7, however, shocked and captivated the world's attention, and the two men were somewhat disappointed by the lack of attention their book received. Reviews of *Sea of Cortez* were mostly favorable, if tepid. Critics such as Clifton Fadiman of the *New Yorker* doubted the book's collaborative authorship, stating, "I think we may safely assume that Mr. Ricketts contributed some of the biology and Mr. Steinbeck all of the prose" (December 6, 1941). The men adamantly defended the text's coauthorship and wrote to publisher Pat Covici in August 1941 that "the structure is a collaboration, but mostly shaped by John" (Astro, Introduction xvi). In December 1941 Ricketts wrote to Campbell expressing his pleasure with the result. "I was very charmed with the book. John certainly built it carefully. The increasing hints towards purity of thinking, then building up toward the center of the book, on Easter Sunday, with the nonteleological essay. The little waves at the start and the little waves at the finish, and the working out of the microcosm-macrocosm thing towards the end. I read it over more than I do lots of other things still" (December 31, 1941). Ricketts's essay "Non-teleological Thinking" was included in its entirety in chapter 14 as the Easter Sunday sermon and became the only philosophical essay of his to be published in his lifetime.

When *Sea of Cortez* was published in December 1941, sales were low, and in the early 1940s Ricketts found himself in financial trouble. The war had diminished orders and the lab's business had fallen off. In a letter to Nan dated March 13, 1942, Ricketts expresses his concern:

"Royalties on the book have so far been nil, and again I am afraid I fathered a financial flop." Needing a break and hoping to begin research for a project based on the outer shores of Vancouver and the Queen Charlottes in Canada, Ricketts and Toni traveled to Hoodsport, Washington, for a few weeks during the summer of 1942. Meanwhile, Steinbeck's marriage to Carol continued to deteriorate. His relationship with Hollywood singer Gwendolyn Conger—whom he met in the summer of 1939—ultimately led to his final separation and divorce from Carol in 1943. Upon his return to California from Washington, Ricketts was drafted as a medical technician at the presidio in Monterey, and he closed Pacific Biological Laboratories for the duration of the war.

In July 1942, Ricketts composed one of several documents regarding information he had gathered through colleagues and research throughout the early 1940s about the Palau (also Palao or Pelew) Islands. This small group of islands, located within the Caroline Islands of the Pacific Ocean, had once been occupied by Germany. During World War I, Japan took possession of the islands, which were officially mandated to the Japanese in 1920. Palau became the site of one of Japan's air bases and a strategic location during World War II. After the bombing of Pearl Harbor, the United States recognized the value in securing the Palau Islands to access the Japanese islands directly. There was, however, little geographical information available about the Palau area—with the exception of scientific documents.

Ricketts believed the U.S. government, specifically the navy, would benefit from the published scientific studies of the Palau region, particularly those concerning the region's topography and ecology. These surveys, conducted primarily by oceanographers and biologists, provide detailed data about coastlines and waters surrounding the islands, including weather, vegetation, water depths, currents, and tides. Such information, Ricketts believed, would be vital for any strategic plan the government might consider. "I have been doing this for myself," he writes, "for my own fun if you could call it that, and as a cure for that curious malady, the war jitters. But if it may be valuable to others, I cannot see why it shouldn't be shared. Furthermore, if there is any truth at all in the old saw about knowledge being power, then the more we know about these things, the more fitted we'll be to fight the good fight and to make philosophically what sacrifices we must take" ("Palao

1942" 1). Ricketts wanted the government to become aware—if it had not already—of the available information and data published by scientists about the Palau islands. He compiled his own research in an outline, "Synonymy of the Palao Islands," which includes a list of synonyms of the islands, an annotated bibliography, and "suggestions regarding additional work along these lines" ("Synonymy" title page); Ricketts hoped to continue his work or pass it along to someone else after his inevitable draft notice arrived.

In the meantime, however, Ricketts contacted military officials and discussed the matter with Steinbeck. On April 28, 1942, he writes to Steinbeck, "You will recall that sometime back I wrote you, and you replied saying that the info had been [forwarded] to the Asst [Secretary] of War, on the info available in reports of the marine biol. of the Jap Mandated Islands." Steinbeck recalls this situation in "About Ed Ricketts." "I wrote to the Secretary of the Navy, at that time the Honorable Frank Knox, again telling the story of the island material. [. . .] And I have always wondered whether they had the information or got it. I wonder whether some of the soldiers whose landing craft grounded a quarter of a mile from the beach and who had to wade ashore under fire had the feeling that bottom and tidal range either were not known or were ignored. I don't know" (270). Officers from naval intelligence did, in fact, visit Ricketts at the lab in mid-1942 but left after a brief interview (though some letters were exchanged later). Both men were disappointed, but Ricketts was not deterred from his investigation of the Palau Islands and continued his research after he was drafted.

While in the army, Ricketts worked at the presidio in Monterey as a lab technician—running blood and urine tests—which allowed him to spend most nights at home. He was able to continue gathering data and study the northern Pacific, on local collecting trips, in preparation for his next project, "The Outer Shores." When he was discharged, Pacific Biological Laboratories was still bringing in little business, so Ricketts went to work for the California Packing Corporation as a chemist. He also conducted studies of sardines, recording population fluctuations and contributing to the yearly sardine supplements published in the *Monterey Peninsula Herald,* which primarily reported annual population statistics for sardines. Ricketts tracked sardines consistently throughout the 1940s, and typescripts of his findings exist in which he speculates upon causes of the sardine population decline. For instance, in a December 4 typescript of his 1946 report, Ricketts states: "[T]his year, with the sardine population perhaps at a record low, we have the greatest number of plants in the history of the industry, with a greater number of larger and more efficient boats than ever before, scouring the ocean more intensely than at any time in the past. [. . .] we can answer for ourselves the question 'What became of the fish': they're in cans. And remedial steps involving conservation can and should be taken" (2, 3). Ricketts's advocation of conservation to slow sardine population decrease went unheeded by fishermen and cannery owners, and the following year he amended his report, including an explanation about the effect of plankton levels on the sardine. He attempted to illustrate more fully the interdependence within the delicate ecosystem of Monterey Bay: "The production of plankton is known to vary enormously from year to year. If feeding conditions are good, the sardines will all be fat, and each will produce great quantities of eggs" ("1947 Sardine" 1). Despite the efforts of Ricketts and others who saw the impact of overfishing, however, fishermen and cannery owners continued to exploit the sardine, leaving it virtually extinct in Monterey Bay by the late 1940s.

Home life in the lab at this time was relatively quiet for Ricketts, Toni, and Kay. Edward Jr. went into the army in March 1943, and Steinbeck had been living in New York City since 1941. Yet if Ricketts's life settled into the ordinary during the mid 1940s, it would not be for long. Steinbeck was to immortalize him and significantly, if briefly, disturb the scientist's peace of mind with the 1945 publication of *Cannery Row*. Upon his return to the United States after a stint as a war correspondent, Steinbeck began writing a "little book" filled with humor and nostalgia for his days in Cannery Row and included undoubtedly his most celebrated fictional portrait of Ed Ricketts. The book is dedicated to "Ed Ricketts who knows why or should."

Steinbeck's Doc is both an accurate and an exaggerated portrait of Ed Ricketts. Doc's interest in philosophy, music, and poetry derives directly from Ricketts's own. Certainly, "[Doc] became the fountain of philosophy and science and art." Steinbeck wrote: "In the laboratory the girls from Dora's heard the Plain Songs and Gregorian music for the first time. Lee Chong listened while Li Po was read to him in English" (28). Ricketts's interest in science, however, was undoubtedly more profound than Steinbeck suggests in his fictional portrait. And, in addition, Ricketts's emotions were more thoroughly under check. When Mack and the boys throw a party at the lab that leads to its destruction, "Doc's small hard fist whipped out and splashed against Mack's mouth. Doc's eyes shone with a red animal rage. Mack sat down heavily on the floor. Doc's fist was hard and sharp" (130). Ricketts was known for his even temper, and even in the most extreme circumstances he did not resort to violence, as evidenced in the following example of an encounter with an aggressive *Cannery Row* fan:

> We had last Sunday the only really bad Cannery Row experience to date. It was pleasant noon, I was sitting barefoot and with nothing on otherwise but [a] shirt drinking coffee, reading. Toni just got up and was reaching for her dressing gown. Kay going out had left the door unlocked, which we're careful not to do Sat. or Sunday; there are tourists. So this god dam fellow came in saying not a word, started to walk in our bedroom. I said how do you do [and] blocking the way asked him if he was looking for someone. Toni got behind the doorway. Said he wanted to see Doc, that his wife dared him to come in. With some actual pushing I headed him out into the office, closed our door and he still had the nerve—tho I told him now this was pretty much of a private dwelling—to say he'd like to bring his wife in. At which here she was coming up the stairs. Ended up that I actually pushed him out the door against his wife, shut and locked it in his face with as much nonviolence as I could. Well I didn't get angry, never did feel even actually unkind. But it proves something or other alright. His last words were "wait a minute, won't you tell me what you did with all those frogs, those tom cats; bothered by a lot of people coming in you ought to charge admission." The others have been good people—very much moved by their own inward most kind projections onto what they think I am. And they certainly merit gentleness. But this guy was what you call unsavory at least. (April 9, 1946)

The success of the book, published in 1945, brought Ricketts much unwelcome attention, and Cannery Row itself became a tourist attraction of sorts. However, Ricketts maintained a good-natured attitude

1 / *Biographical Essay*

throughout the constant interruptions and inquiries made by Steinbeck's (and now his) fans.

A more emotional situation at home was compounding Ricketts's frustration about unwanted guests at the lab. Kay, who had been diagnosed in the early 1940s with a brain tumor, was steadily worsening during 1946 and 1947. Toni began to spend a great deal of time with her in the hospital, seeing Ricketts only on weekends and when Kay was able to be at home with them. In addition to caring for Kay and Toni, Ricketts spent increasing amounts of time on his plans for the "Outer Shores" book and applied for a Guggenheim Fellowship to finance an extensive collecting trip in 1948. In a typescript of his Guggenheim proposal, Ricketts outlined his project as twofold: "[First,] to make available to laymen, students and scientific workers detailed ecological and biological information on the great faunal provinces of the Pacific Coast from Panama to the Bering Sea, each province to be considered in a separate volume. [. . . Second,] construct a manual of the invertebrates of the entire Pacific Coast, which will summarize the cataloging and identification aspects of the previous works" (1, 2). Ricketts and Toni had traveled to Vancouver Island and the Queen Charlottes during the summers of 1945 and 1946, and Ricketts was eager to complete what he considered the third book of the trilogy consisting of *Between Pacific Tides*, *Sea of Cortez*, and "The Outer Shores"—ecological texts accounting for virtually the entire Pacific Coast of North America. In underscoring the significance of his work, Ricketts stated he would

> cast light on the laws governing the distributions of animals and . . . contribute towards new zoo-geographical and ecological concepts regarding the mutual relations between the animals and their environments. The ecological approach leads into a field defined by Gislen as "marine sociology" in which the ideas of integration and biological cooperation are implicit. I am at one with Dr. Allee who states, in rating these concepts among the most significant biological developments of recent years: "It is my hope that from the work described . . . all social action may have a somewhat broader and more intelligent foundation." ("Guggenheim" 2)

Despite his time and effort spent on the proposal—and despite a letter of recommendation from Steinbeck—Ricketts was turned down for the Guggenheim. Soon thereafter, on October 5, 1947, Kay died. Grieving, Toni left Ricketts and Cannery Row, moving to southern California with marine biologist Ben Volcani. The Volcanis married and moved to Palestine in 1948.

While certainly melancholy, Ricketts's letters from the months following Kay's death and Toni's departure in October reflect acceptance. He continued to work and carry on with daily activities, but he was also contemplative about the recent events. In a letter to Toni written on December 24, 1947, he expresses his love for her quite objectively, "I'm loving you as deeply and perhaps more clearly than I ever did before." Yet Ricketts was not alone for long. In late 1947 he met Alice Campbell, a twenty-five-year-old philosophy and music major at University of California, Berkeley, and the two married in January 1948. Most of Ricketts's friends on Cannery Row did not know Alice well, and she was not his legal wife, since he and Nan had never finalized their divorce, although Ricketts thought they had. Edward F. Ricketts Jr. remembers little about Alice but does recall her love of Mozart's *Don Giovanni*, which she would listen to on the phonograph while simultaneously reading the sheet music (Interview July 12, 2001). Significantly, Ricketts continued corresponding with Toni until her departure for Palestine in early 1948, and his letters reveal his nonteleological perspective on their situation. In a poignant letter he wrote, "Yesterday I was so tired from no sleep, and I had the curious sadness of knowing then that you were off to Palestine, and [an] allergy I hadn't had for a long time, that it all piled up to I guess the worst choking spell I ever had. [. . .] I got that sadness of things that aren't anymore; as when I talk to Kay in the still same back room (I've had no time to change it around). Es war nicht gewesen sein. It was not meant to be. [. . .] All potential can't become reality. You've got to select. But it makes you sad" (February 4, 1948). While he grieved for Kay, Ricketts continued to plan for his next book, arranging to travel with Alice to the Queen Charlottes in May. In a letter to Joseph Campbell he reveals that Steinbeck, whose marriage to Gwyn had become strained, was once again to be a collaborator on the project and would write the narrative portion of the book from Ricketts's notes.

"We'll be going again to the Queen Charlottes end of May. John is coming up there for part of July. I have turned over to him verbatim transcriptions of my two summer's notes; then he'll have his own and mine for the coming trip. Should be able to construct quite a book out of them; he'll have his journal done I fear far before my scientific part's complete. It should be a smaller *Sea of Cortez*. 'The Outer Shores'" (April 26, 1948). Ricketts had, in fact, prepared his existing notes for Steinbeck. On the last page of his notebook, which he planned to turn over to his friend, he wrote, "well jnny boy this is it, this is 30, the trips of 1945 and 46 are over, its yr bk now and god bless you" ("New Series, Four" 143). In the above letter to Campbell, Ricketts also discusses his desire to reapply for a Guggenheim and Toni's departure for Palestine. The tone of this, one of his final letters, is optimistic and content.

On May 8, 1948, while driving across the railroad tracks on Drake Street off Cannery Row, Ricketts was struck by the Del Monte Express. He was conscious when the men pulled him from his car, telling them, "Don't blame the motorman." He was admitted to the hospital, where he was surrounded by loved ones and friends. In his essay "About Ed Ricketts," Steinbeck wrote:

> The next morning Ed was conscious but very tired and groggy from ether and morphine. His eyes were washed out and he spoke with great difficulty. But he did repeat his first question.
> "How bad is it?"
> The doctor who was in the room caught himself just as he was going to say some soothing nonsense, remembering that Ed was his friend and that Ed loved true things and knew a lot of true things too, so the doctor said, "Very bad."
> Ed didn't ask again. He hung on for a couple of days because his vitality was very great. In fact he hung on so long that some of the doctors began to believe the things they had said about miracles when they knew such a chance to be nonsense. They noted a stronger heartbeat. They saw improved color in his cheeks below the bandages. Ed hung on so long that some people from the waiting room dared to go home to get some sleep.
> And then, as happens so often with men of large vitality, the

energy and the color and the pulse and the breathing went away silently and quickly, and he died. (227)

Steinbeck was in New York at the time and frantically attempted to reach Ricketts in time to say goodbye. "The greatest man in the world is dying and there is nothing I can do," he told his friend Nat Benchley (Benson 615). By the time Steinbeck reached California late in the evening on May 11, Ricketts was dead.

The day after Ricketts's funeral, Steinbeck went to the lab to sort through his friend's personal belongings with another friend, George Robinson. He saved the manuscripts and some of the notebooks Ricketts had kept but burned all of their correspondence. While going through the notebooks, Steinbeck tore out sections and passages he felt were too private for others to see. On the last page of the final notebook Ricketts kept—containing much of the preparatory material he had assembled for the "Outer Shores" project—a grieving Steinbeck scrawled, "E.F.R. was hit by train May 9. Died May 11 1947 J.S." ("New Series, Four" 261). Much of Ricketts's property was given to friends, and Alice, who soon learned that she had not been Ricketts's legal wife, left Monterey altogether.

Steinbeck was devastated over Ricketts's death. He wrote about his grief in letters to friends and eventually paid homage to his friend in "About Ed Ricketts," included as a preface to the 1951 publication of *The Log from the* Sea of Cortez. Indeed, Ricketts did not fade from the memories of those who knew him. "No one who knew him will deny the force and influence of Ed Ricketts," Steinbeck wrote. "Everyone near him was influenced by him, deeply and permanently. Some he taught how to think, others how to see or hear. [. . .] He taught everyone without seeming to" (228).

Pacific Biological Laboratories remains a fixture on Cannery Row, and tourists still come to see this legendary landmark in "Steinbeck Country." Ricketts's landmark research on the Pacific coastline continues to be revered by scientists; *Between Pacific Tides* is presently in its fifth edition; and *The Log from the* Sea of Cortez remains in print and sells well worldwide. Ricketts's specimens are on display at the library at Hopkins Marine Station in Monterey, at the Monterey Bay Aquarium, and at San José State University. More than two hundred specimens at the California Academy of Sciences in San Francisco continue to be used for research today. Stanford University houses Ricketts's letters, and exhibits throughout California devoted to Ricketts draw students and fans fascinated by the legendary Doc of *Cannery Row*. While his influence on Steinbeck alone makes Ricketts a significant figure in literary history, his own writings—his letters in particular—merit attention that is long overdue.

Notes

1. Edward F. Ricketts Jr. believes this letter may not have been sent to Hedgpeth, which seems likely. In *The Outer Shores* he does not mention Strong's letter, asserting Ricketts and Steinbeck met at Calvin's home.

2. Yet in 1978 biologist Joel Hedgpeth edited a collection of Ricketts's essays and notebooks into a two-volume work he aptly titled *The Outer Shores*.

3. Ricketts's essays were published in Hedgpeth's *The Outer Shores* (1978). Unless otherwise indicated, the pages referenced in this essay refer to Ricketts's own writing in Hedgpeth's publication.

Renaissance Man of Cannery Row

1936–1938

On November 25, 1936, a power surge in the Del Mar Cannery caused a fire on Cannery Row that destroyed Pacific Biological Laboratories, including all of Ricketts's personal and professional records and correspondence. Ricketts escaped, saving only his car, the clothes he was wearing, and a portrait of him by James Fitzgerald (see letter dated March 31, 1937; in "About Ed Ricketts," Steinbeck asserts Ricketts also saved his typewriter, but in *The Outer Shores*, Hedgpeth notes that Ricketts's own inventory of lost property did not include his typewriter or car). While rebuilding the lab, Ricketts stayed with Fred and Frances Strong at their home in Pacific Grove. Friends and family members surprised Ricketts at Christmas by giving him some of his favorite books in order to help rebuild his lost library. His gratitude is apparent in his letters.

The following is the earliest letter found after the fire.

To V. E. Bogard and Austin Flanders
December 16, 1936

Dear Bogard and Austin: You will be interested to hear what's been happening.

We have the lot almost entirely cleaned up, still one large pile of wood and debris stacked together. Most of the concrete seems to be alright.

Rough plans have been drawn for a shack-like board and bat structure to house office and labs, 28 x 33'. Estimated cost, top $981, plus $100 for rough in plumbing and $75 for conduit, or vice versa. Labor and material plus 5%. Estimate by builder who does much of the cannery work and who was sent up into Oregon by 3 of the local canneries to build 5 plants there. We have saved lots of conduit, cast iron pipe (all of which is good), and some valves and galvanized pipe that seems alright; this will reduce the cost. Galv. iron too expensive.

As soon as insurance company pays up (they are investigating criminal carelessness) I will have this part erected. Then Roy and I will get busy with 2 x 4s and erect galvanized iron roof over shark tanks, run the conduit and water pipe out there for immediate start on shark preparing.

The balance of the $3000 will have to go for equipment, stationary, such few items of stock as we have to buy (maybe none, can trade for most), a bit to Rodriguez, and small truck to replace present car.

It looks not impossible now, altho still terribly difficult. Very hard to pack orders, without any decent facilities, out in the rain.

Insurance company investigation has been delegated to [a] local electrical engineer, and I have given him [a] statement. He has worked out a very interesting picture, amply supported by evidence. There were about 4 pieces of neglect, any one of which might now have resulted thus, but the combination of which was sure to cause terrible fire sooner or later. He says the evidence, from the type of wiring, fact that fuses didn't blow as they should etc (on the power pole), is that the full power load of between 500 and 1000 horsepower at 2300 volts was flowing both thru our place and the entire Del Mar plant, that all the wires took afire and burned until they burned out, starting dozens of tremendously hot fires. The fire is known now to have started at the main switchboard that fed all 5 of the Del Mar Canning Co. plants. He says that ordinary house fuses are no valve in the face of such a load; they blow but still transmit. I'm glad I didn't go near our wiring. Next time I'm going to try to devise some protection against this thing that the electric companies say can't happen. I think a 5 or 10′ length of smaller diameter, before the service comes to the switchboard, would surely do it. That would burn before the inside or switchboard wires took afire.

I am still staying at Frances', Carmel 1146 in case you need to phone. Soon as roof is up again at the lab I'll move back down there. The cannery is *not* going to rebuild this year, despite previous reports, and there's no certainty that they'll ever rebuild here. The city seems to be awfully mad at them, and they surely should be.

V. E. Bogard (1889–1952) worked at University Apparatus Company, which sold laboratory supplies and equipment primarily to the University of California, Berkeley. Bogard worked closely with Austin Flanders, a distant maternal relative of Ricketts. According to Nan Ricketts, "Austin was a major stockholder and president of University Apparatus Co. He had lived in Berkeley for many years, and knew the state well, and all the interesting places" ("Recollections" 26). Flanders helped the Rickettses settle into California during the 1920s, taking them on trips to Santa Cruz and Yosemite and showing them local areas of interest.

In 1924, Austin Flanders became a shareholder in Pacific Biological Laboratories. As part of the agreement, University Apparatus Company paid off the lab's debt, and Flanders was named vice president of Pacific Biological Laboratories.

The "criminal carelessness" Ricketts mentions did, in fact, lead to a lawsuit: *Ricketts and the Del Mar Cannery* vs. *Pacific Gas and Electric*. Pacific Gas and Electric won. In a 1937 letter to Jack Calvin, Ricketts comments on the waterfront fire and loss of Pacific Biological Laboratories: "Insurance totaled only $3000; loss was nearly $12,000." While Ricketts did, in fact, collect the three thousand dollars insurance money, he never received additional reimbursement from the lawsuit.

Roy Lehman worked intermittently for Ricketts at Pacific Biological Laboratories assisting with lab work and collecting. Vincent Rodriguez owned much of the property on Cannery Row and had sold Ricketts the lot on which Pacific Biological Laboratories stood.

~

To Colonel and Mrs. Wood
Pacific Grove, California
December 26, 1936

Dear Col. and Mrs. Wood:
The candy arrived for the children and they were very thrilled—the best by far they have ever had, and the biggest box: to which chil-

dren aren't insensible. Part of your good wine already has been enjoyed under the proper auspices of good fellowship and good food.

As John may have told you, I was very much surprised Christmas Eve at the Christmas tree celebration, preparations for which he knew about, as everyone else did. But I didn't. When I was asked to pass out a group of wrapped presents—presumably to little Peter— I was surprised to find the first addressed to me. Put it aside, with modesty presumably becoming. The next also was to me, and the third. Then I tumbled and almost wept. Odyssey, a fine German *Faust*, Blake (facsimiles of the original illuminated copy), 4 of Jeffers, the new edition of the World Poetry Anthology, Shelley and Keats, a fine Walt Whitman, a new and good translation of the tao of Lao Tse, other things. I put "Darkling Plain" with these things where it doubly belonged: in that company, and because it too was in the nature of a rehabilitation present at Christmas time. Have read already much of it; I read slowly and not often. I like an exact and lovely economy of words. As in many of those things, but I noted particularly "A Neighbor differs with the Master of Tor House." Fine scientific exactness: "earlier darkness departed later." The fine honesty, really quite deep, of a thing found to be "so" (whatever the condition may be), and then calmly and quietly stated so with exact economy of words, hasn't been considered often enough. Leaving the words, I have thought a good deal about the contents of that poem; I mean even before I read it, because that's a constant problem. I worked up some ideas of it once. John has a copy of the essay-like thing that resulted. It considered Jeffers. When (if!) I can get to making another copy, I'll mail it on. Think you'll be interested.

Reading this, I hope her physician will suggest the oil for your grandchild. Attached is the information I have on shark liver oil. I hope it fills the bill. I am wishing a most happy New Year to you.

Ed

Col. Charles Erskine Scott Wood (1852–1944) managed three distinct careers in his lifetime: as an infantry officer in the Indian campaigns; as an attorney in Portland, Oregon; and as a poet and satirist in Los Gatos, California. From his home, The Cats, which he shared with his second wife, poet Sara Bard Field (1883–1974), Wood maintained friendships with renowned artists such as Ansel Adams, Robinson Jeffers, and John Steinbeck, among others. Wood and Ricketts most likely met through Steinbeck, as both Wood and Steinbeck were living in Los Gatos at the time.

Darkling Plain was Sara Bard Field's 1936 volume of poetry published by Random House. As Ricketts indicates in his letter, Field gave him a copy soon after the lab fire.

The following was attached to the Woods letter dated December 26, 1936.

To Colonel and Mrs. Wood
Pacific Grove, California
December 27, 1936

I heard of shark liver oil first through G. E. MacGinitie, then associate professor at the Hopkins Marine Station (of Stanford University) here at Pacific Grove. He said it represented an entirely new principle, that their professor of biochemistry, Dr. Hashimoto, had reported on it verbally at a seminar, after having done quite a bit of work on it. Mac was giving it to his son Walter. Later it came to be known quite favorably locally for its general tonic effects, and especially in arthritis and allergic conditions. How it was originally discovered I cannot imagine, although even that might still be traced, since the reduction plant operator who originally produced the oil is still here.

Shark liver oil of this type is not a vitamin concentrate. A number of analyses and evaluations have been made, and the vitamin content is known to be low, almost negligible. Some other factor operates. Hashimoto stated that the unknown active element or elements simulated cortin (extract of the cortical layer of the supra renal, the core of which secrets adrenaline) a vitally necessary hormone-like substance which is being used experimentally to keep alive cats that have been adrenalectomized. I cannot instantly check up on cortin, my reference library is gone without trace and "the place thereof knows it no more," but as I recall, it's vitally important now in physiological research and gland therapy, expensive, and difficult to obtain in even reasonable purity.

Anyway, vitamins are known not to be present, at least not in significant concentration, and the usually spectacular results have to be attributed elsewhere. I think of the situation symbolically in the following light:

During certain illnesses, the normal vitamin intake of the human body seems inadequate, and enormous concentrations have to be prescribed. It is as though the vitamin retaining (or elaborating or utilizing) mechanism were a sieve, with interstices small during health so that a high proportion of the vitamins normally present in food would be retained, but with the mesh becoming large and loose in certain pathological conditions so that to get the same amount of total retention, enormous dosages would be required. (Of course the situation probably isn't mechanical at all, and it's factually incorrect to consider it with reference to be at least symbolically enlightening). Now carrying on this confessedly inadequate figure of speech, it is as though shark liver oil contracted to normal proportions the vitamin retaining mesh, so that the normal intake of vitamins became again adequate.

That's a bad example and a bad hypothesis despite its aptness, but it often conveys an emotional understanding of a situation quite different than pictured, and not picturable otherwise.

Since Dr. Hashimoto's original research at Hopkins Marine Station, either he severed his relations with Stanford or they with him. He left without publishing, went to Monterey Hospital on a research fellowship endowed by Mrs. Sidney Fish (for the purpose of continuing s.l.o. [shark liver oil] investigation I understand) but again left no notes and published nothing, going from there to UC (I think in the medical school at S.F.). Now he is in Los Angeles on his own, still working with the oil, according to Schaeffer who sends him 5 gallons of the commercially produced oil from time to time. Schaeffer says that Hashimoto is working out a method of concentrating the aphrodisiacal qualities of the oil for the benefit of Hollywoodites.

I had in my files, now burned, letters from several hospitals, clinics, and individuals who commented on the successful use of this oil in conditions of post-operative shock following removal of ovaries, etc. There is abundant evidence of its efficacy in arthritis and asthma. Nothing of course can be done with it in a large way until the active ingredients have been isolated and studied—a research problem of considerable magnitude—but in the meantime, there's no reason why individuals can't profit by its use. I know dozens of people who have taken it just as they would take cod-liver oils and there's been not a single untoward result. Only one case, Isabelle West, of no effect. Experience indicates there's no need of its constant use. My mother took it for a few months only, several years back until she recovered completely from her arthritis, and hasn't touched it since. When any of us get run down, suffer from recurrent colds, we take shark liver oil (if it's available) for a month or so.

Availability is the greatest difficulty. The great basking sharks from which this oil is produced can be taken here only at certain seasons of the year, and sometimes they're missing for a year or more. The capture of so large an animal offers difficulties in the way of equipment and technique. Whale handling methods would probably work. At present no one fishes them. Frank Lloyd and Jo Machado took a few, then Frank Lloyd and Hilary Belloc. But Hilary has gone up to SF to work in the book department of the Emporium, and anyway, as Frank says who has gone back to fishing rock cod—there aren't any sharks and we lose too many harpoons.

When the properties of this oil become generally known and accepted, the large demand will automatically take care of the supply situation, boats will be especially equipped, and fishermen from here north will be capturing these cosmopolitan north-temperate-zone fish. Until then, we'll just have to make out with what oil is available, worrying more about supply than about rendering methods.

↩

To Thayer Ricketts
February 8, 1937

Dear Thayer and Evelyn:
There has been little time for writing you since the fire. Hasn't this been a trouble time for the Ricketts family! Everyone seems to have been hit. You should have seen me roaming around early in the morning, shivering without shoes or sox, or even adequate clothing, looking for hot coffee in places not even open yet. Not much shock to it otherwise, tho. I expected not to be able to sleep for seeing the flames leaping up, but it didn't bother me at all. Altho the last time in SF, at a cheap hotel, I couldn't sleep until I had made sure I knew the way to the fire escape, and that my flashlight was at hand. Amazing how fast a thing like that can go. But apparently 2300 volts at 1500 horsepower wasn't being pumped thru the cannery, and possibly into our plant.

It was good and kind of you folks sending me that nice pen and pencil—at a time particularly when things haven't been too easy with you. I got a grand start at the shower Frances arranged; I was very much surprised. About 20 books, clothes, etc. John sent electric coffee maker. 8 cup size. I thought how appropriate it was for him. Enough coffee for him and me during one of our long confabs.

I thanked our lucky stars that you hadn't already assembled the glass sealing outfit before the fire. It would have been gone too, with nothing to show. Keep it in mind still. We want and need it now more than ever before. If I can get a bulge on even our few competitors in this not-highly-competitive field, it will be particularly nice from now on. We need every advantage we can get. Hard times ahead. I can envision a vial-package that will be more serviceable than anything heretofore available, and that will make a big hit with ye goodly prof. With the strain that has been put on you financially and in a business sense recently—Uncle Gwyn sick etc—I imagine you won't have even a minute to work on such a thing for some time, but keep it in the back of your mind anyway. There may sometime soon be an opportunity. I can take care of any required expenses; just let me know beforehand. Another thing I thought to mention: We have in the past, and will again in the future, do a lot of delicate operating on cats in connection with embalming. We use a good many dozen cats in a year, and can use more as I develop better sources of supply. Adequate shadowless light is always problem. If you ever take in thru exchange, any junky Operay lights—the overhead type—that are inadequate for hospital use and so have little value, we might be able to pick one up at low cost. We wouldn't be justified in paying much; in the past we used a bank of 5 drop cords with green porcelain shades, and that was OK. But a shadowless light up out of the way would be convenient if it could be got cheaply. Don't know if ever such an opportunity will be available; I may be barking up the wrong tree. Maybe you can use all 2nd hand equipment that comes in there. I can always dig up five or ten dollars for some useful gadget.

Looks as tho the firm will be able to pull out of this, and I also. But it's a great strain. I have not only more work than I can hope to do, but much more than even an efficient person could figure on doing. The total actual loss must be more than $12,000; and there was only $3,000 insurance! Then there were a thousand and one little personal things destroyed that I hate to lose. A good many were superfluous, but some are essential to proper conduct of the business and/or my comfort. If everything is going along after a year more, we shall have survived, but it will be a struggle.

We are still looking forward to your visit here, and I hope eventually you can make it. I'll even have things fixed up soon so that a couple of guests can be put up at the lab—under strange but quite convenient and pleasant conditions. I'm staying here now. It's pretty comfortable. A Franklin stove (fireplace type) went in this afternoon, and there is a floor furnace. Shower bath, gas stove, even an indoor toilet—a luxury not possible before. I have really a pretty nice apartment, and an extra room was built that can be used as a spare lab or as a spare bedroom. All board and bat construction, but my room and the office have been insulated with 3 ply and celotex and are really pretty comfortable.

Well, this is the 5th personal letter today—and me with government reports to get out. Those periodical reports are the bane of my existence. Sales tax, employee's security, alcohol report; like feeding white rats or shaving, a job that's never done. Say hello to the folks for us.

Ed

Thayer Ricketts (1902–ca. 1988), youngest of the three Ricketts children, worked for Scanlan Laboratories, which manufactured surgical equipment. Of Thayer's relationship with Ricketts, Edward F. Ricketts Jr. wrote, "I always had the impression that brother Thayer and EFR weren't especially fond of each other" (Ricketts E-mail, October 3, 2000).

To Howard and Emma Flanders
February 8, 1937

Dear Howard and Emma
Frances probably wrote you the results of the Christmas surprise. I am just now getting to write letters that should have followed that good thing more immediately. But trying to reconstruct 15 years work immediately, with not much money, and with the pressure of schools demanding their materials, of course is a big job. And I'm not so big; so there's sure to be a discrepancy.

That edition of Whitman was a happy thought. I shouldn't have remained very long without *Leaves of Grass* in any case, but wouldn't have felt justified in getting that very good edition, which is by the way my favorite. So it represents a luxury I couldn't otherwise have had. I suppose you've seen it. Many of the Rockwell Kent [drawings] are fine. He's pretty lusty too. Like Whitman.

Already, just from that rehabilitation shower, I have the foundations of a new library. Not least, but best of all, is the glow I got out of knowing there were so many good people. I was certainly surprised. Trouble is good. It shows things up.

The lab is pretty well put back together. I was surprised to see how much could be done with a little money, some common sense, and a bit of accumulated experience. Some, at least, of the previous errors weren't reenacted. Funny, tho, how hard it is to start out. I'd (almost literally) reach for a sheet of paper, or for a well accustomed reference, only to find that it no longer existed, and that even the place where it stood was gone. Imagine me (at first) prowling around, cold and wanting hot coffee; no shoes or sox, or coat. There was no point of reference aside from a blazing heap of ashes and a car. Now all fine again though, and if, within another year, the firm shall have weathered the storm, and I shall have also, it will have been a good thing. As I suppose everything that happens is a good thing, if a person looks at it that way. Schools and business associates have been very cooperative; orders are going out again already—in fact we kept right up in a limited way. But wotta mess!

I seemed to have missed you folks the last time or two. When you come down again, get in touch with me soon (before you start if possible) if you have no better place to stay. I'll be fixed here within another couple of weeks so I can put up two guests. Fairly well too. You'll be most welcome. I'm glad to have people pop in. Good stunt to bring blankets. If the weather is as it has been, bring along fur coats too!

Thanks again to you.

> Howard and Emma Flanders were distant maternal relatives with whom Ricketts maintained a somewhat regular correspondence.

Howard was a general practitioner who sometimes sent specimens to Ricketts; in July 1938, he sent Ricketts a human embryo, which Ricketts hoped to dissect and preserve for further study.

Rockwell Kent (1882–1971) was an artist whose home on Monhegan Island off of the coast of Maine was later bought by James Fitzgerald. Kent was a left-wing activist blacklisted during the McCarthy hearings and was associated with the social realist Ash Can artists. While he was known primarily as a painter, Kent also established himself as an engraver and lithographer.

To Austin and Hazel Flanders
February 8, 1937

Dear Austin and Hazel:
I am just now getting around to writing letters about the Christmas shower—my start at rehabilitation. Seems a long time, but at first I was mixed up. Then when building started and orders again were being attended to, there wasn't time. As there still isn't! But it's nice to write anyway.

Mother was glad to go up there; I know she's having a good time. Wonder if Punk will complicate matters. She made some very attractive curtains for the new office and for my living room and library before leaving.

I hope you can drive down here soon. You'll be surprised at how much has been done in a short time and with a little money. The notion of starting out with a cheap shack was a good idea. That was the first light I saw in a complicated situation. If I had started out with the idea of trying to put up a good building, hung between cement walls, as I should have liked to do, it would have been impossible right from the start. As it is, a pretty good building eventuated. We succeeded in paying off $500 of the mortgage out of the insurance money too. And I turned in the Packard on a Ford. A funny little car. V-8 60 h.p. Smallest standard car built I imagine. I don't like it very well, but it will provide cheap transportation. I presume there will never be so long cost per ton-hour-mile as in that old Packard, but now I'm trying to avoid both the miles and the tons. Since there is no longer a big family to cart around on long collecting trips, nor for that matter, not even the possibility of very long or time-consuming trip, this should be easy.

Sharks coming in, being prepared, and going right out as if there hadn't been any fire. Looks as tho the firm will be able to pull thru that disaster, and if a projected law suit against the PG&E (vs insurance company, cannery and PBL) works out, it won't be bad at all. Insurance company lawyers will handle it. They claim they have the thing dead to rights, so much so that the local representative thinks our end of it can be handled on a contingent basis. But the inventory is involving a tremendous lot of work, so that I have to neglect too many other things!

Well, mother will have the rest of the news for you. I may be able to get up there to bring her back; will figure on doing so, and will plan on coming so as to be able to visit there a little. The first time literally in years! Otherwise I'll hope to see you here soon.

The Blake book, by the way, that you folks provided for through the goodness of your heart, and certainly all unbeknownst to me until it happened, is a luxury I had been thinking about even before the fire, but one I never expected to have. That one evening started things going for me again in the way of clothes and library. The amount of kindness turned out by the fire almost justified it. The two best suits Austin brought down I have had remade and you'd be surprised how swank they are. I haven't been better dressed since I came to California.

> Austin and Hazel Flanders were also maternal relatives with whom Ricketts corresponded regularly.

To John and Alice Cohee
February 10, 1937

Dear Jon & Alice:
Much delayed, here are the transcripts of the two essays I mentioned to you a couple of months back. Jim [Fitzgerald] has one copy, and I never did yet get it back from him—altho seeing him frequently. These came from Dick Albee thru Paul Budd.

Typing isn't very legible, but it can be read; I'm sure you'll find it worth while. Fun anyway. Two or three other people are on the waiting list, and now, since the fire, these are my only copies, so figure on shooting them back within a couple of weeks or so when you are thru with them.

As you can imagine, I've been pretty busy. The lab is put back together after a fashion, but much remains still to be done. Will take years to equip it as it was. But in the meantime I have pretty pleasant living quarters if nothing else. Eventually too I'll have space for an extra person or two, under conditions involving a strange, but not unpleasant environment. Come along; let me know beforehand preferably; someone probably will have to sleep on the floor but "think nothing of it." I'll see you if and when I get up there, but nobody knows when that'll be, and won't tell.

Ed.

 John and Alice Cohee were Ricketts's friends with whom he corresponded periodically.
 The two essays mentioned in the first paragraph were likely two of Ricketts's three philosophical essays: "Non-teleological Thinking," "The Philosophy of 'Breaking Through,'" and "The Spiritual Morphology of Poetry," which circulated among his friends throughout the 1930s and 1940s.

Ricketts's fascination with poetry, which he explores in "The Spiritual Morphology of Poetry," gave rise to his desire to read foreign verse in its original language—even if he did not know that language fluently. He spent hours studying German in order to read *Faust* in its truest form. The following are just two of many similar letters requesting information about foreign poems.

To Department of German
University of California, Berkeley
February 15, 1937

Dept. of German,
University of California,
Berkeley, Calif.

Gentlemen:

Can you give me readily the German title of, and a bibliographic reference to, the poem of Justinus Kerner which appears in the translation on p. 855 in the Mark Van Doren Anthology of World Poetry, as "Homesickness?" Once I searched there at UC main library for the original in Kerner's works, but lacking the title, was not successful, and it occurred to me that a few words from you might save me additional possibly fruitless effort.
          Sincerely,
          E. F. Ricketts

To Department of Spanish
University of California, Berkeley
February 15, 1937

Dept. of Spanish,
University of California,
Berkeley, Calif.

Gentlemen:
Can you give me readily the Spanish title of, and a bibliographic reference to, the poem by Pedro Calderon de la Barca which appears

in translation on p. 645 in the Mark Van Doren Anthology of World Poetry, as "The Dream Called Life?" Once I searched there at UC main library for the original in Calderon's plays, especially in "La Vida es Sueno," but had no luck, and it occurred to me that a few words from you as to original title etc, might save me additional possibly fruitless effort.

Sincerely,
E. F. Ricketts

To Monterey County Hospital
Salinas, California
February 19, 1937

County Hospital
Salinas, Calif.

Gentlemen:
A cannery worker, an independent and industrious man past middle age, lives next door to the laboratory here, in a tent in an empty lot. He has been very kind and cooperative to me personally, in the hectic times that followed the disastrous waterfront fire here and which destroyed all our assets, and in the rebuilding days. He has a pathological eye that needs attention. I presume he is eligible for county help, since he has been living here at Monterey for some time. I feel quite grateful to him, and would be willing to see that he got over that cost free as to transportation, if you are willing to have a look at the trouble. If this can be arranged, I shall be glad to know which days of the week are most suitable from your standpoint.

Sincerely,
E. F. Ricketts

The cannery worker mentioned in the preceding letter was most likely Harold "Gabe" Bicknell, who later inspired the character Mack in Steinbeck's *Cannery Row*. Edward F. Ricketts Jr. had this to say:

"I'll bet the person referred to in the County Hospital letter was Gabe. I knew he had a weird eye, but I thot it was the result of syphilis" (Ricketts E-mail, October 11, 2000).

To Torsten Gislen
March 31, 1937

Dear Gislen:
Chief purpose—hasty note—is to say that John Steinbeck is enroute to NY, Sweden, and Russia; may have a chance to see you, and I have urged him to look you up. I am taking care of his dog during the six months. He figured on being in north Europe, probably Denmark, for mid-summer night. He has read some of your papers—"Tendencies toward death and renewal" at any rate. He is a good man, now a famous author. I hope he likes you and you him. His "Of Mice and Men" which followed "In Dubious Battle" I considered a fine piece of work. Classed often more or less as a mystic; I think not, except in the finer sense of the word. Certainly unconventional, formality would surely call him bohemian.

Many things have happened since I last wrote. On Nov. 25th I had the champion of all strokes of bad luck. During the Monterey waterfront fire, the lab was completely destroyed. For sometime now completely separated from Nan, I was living here. Sleeping at the time. Got out with my life and that's about all; had time to put on pair of pants and shirt. Succeeded in getting the car out, and a painting of which I was fond—James Fitzgerald's portrait of me. My library, general and scientific, all my personal belongings, clothes, furniture, including a few pieces that had been in the family for more than a hundred years and which came to me at Father's death last year, all destroyed. Most of the records, except for some in the safe, are gone, and with them records of distribution and methods. Worst of all, my library had been assuming good proportions, possibly the best on the Pacific coast with reference to local marine ecology. In this connection, if you have still available any copies of your Misaki,

Gullmar Fjord and "Tendencies towards death and renewal" papers, I'll be exceedingly glad to have them. I suppose too late now. I had amassed really a good library on Pacific coast invertebrates, including the Carlgren anemone papers, and several on polyclads and enteropnesuts. Many I can never replace. And only a little insurance, $3,000. Loss 4 or 5 times that. Not enough even to replace the buildings, but I am going ahead again with what little there is; if I can keep up the energy and interest, plus a little luck, I can get back again.

Nothing recent from Stanford Press about the book. It has been scheduled for issuance a dozen times during the past few years, but they are slow. Our contract reads "Within a reasonable time." Best of regards to you and the family.

Torsten Gislen (1893–1954) was a Swedish marine biologist, botanist, writer, and close family friend with whom Ricketts corresponded about personal and professional matters. In her memoir, "Recollections," Nan Ricketts writes that Dr. Gislen was from "Lund University in Oslo. [. . .] The Gislens were in this country for two years, and I believe they spent most of that time on the Monterey Peninsula. They lived in a rented house in Pacific Grove and we saw much of them and took them on many trips" (23, 24).

John and Carol Steinbeck left California on March 23, 1937, to travel to the East Coast and then onto Europe. The success of both *Tortilla Flat* (1935) and *Of Mice and Men* (1937) gave the Steinbecks the financial means to travel. There is no definite record, however, of a meeting between Steinbeck and Gislen during Steinbeck's stay in Sweden.

The book mentioned in the final paragraph is *Between Pacific Tides,* which was eventually published in 1939 by Stanford University Press.

⌒

Ricketts's interest in art permeates his writing—both in essays and correspondence. In fact, he believed that evidence of breaking through from the physical to the metaphysical was often most apparent in the fine arts, including painting, music, and poetry.

To Museum of Modern Art
New York, New York
April 7, 1937

Museum of Modern Art,
11 West 53rd St.,
New York City, N.Y.

Gentlemen:
Per your lists recently submitted, and upon receipt of the enclosed remittance, will you kindly forward me the following reproductions:

From Leaflet: Modern Art Prints:
Series A—Paul Cezanne. A1 thru A10 .50

From Selected List of Color Postcards, Raymond and Raymond, N.Y. (I assume these refer to color collotype cards at 10¢ each, altho no price is mentioned):

Fra Angelica: 4 as listed
Blake: 2 as listed
Van Eyck: 9 as listed
Giotto: St. Francis etc
El Greco: Agony in the Garden
       Burial of Count Orgas
       Holy Family
       Mater Dolorosa
       Pieta
Holbein the Elder: 2 as listed
Holbein the Younger: 5 as listed
Memling: 16 as listed *one set to Lucile*

Modigliani: Nude
Picasso: as listed except
  for: The Guitarist .10 only
Renoir: Bather, Torso
      Farm on the Seine
      La Loge
      La Premiere Sortie
      Madame Renoir
      Spring
      Meadow Flowers
      Paris Boulevard
      Two Bathers
      Woman Sewing

Rivera: 14 as listed
Sasseta: St. Francis and the Wolf
Vermeer: 8 as listed   88 color cards at .10 ea 8.80

                                                       9.30

                      Sincerely
                      E. F. Ricketts
                      Pacific Biological Laboratories
                      Pacific Grove, Calif.

To Dick and Jan Albee
June 8, 1937

Dick and Jan, Dk & Jn, D & J Albee
979 Rhode Island Ave.,
San Francisco, Calif.
Tuesday

What letter? Suppose I didn't write no letter. Suppose I wasn't even here—that I was in SF attending the 1938 World's Fair. Or that there shouldn't be any World's Fair, or any 1938, or even any world. Maybe it all happened in the future and this is the scatter effect of Seashore's psychological notion—ripples. Paul Budd that rat is off for Hawaii. Maybe he doesn't like our fog. And Gret [Gretchen Schoeninger] to Gaviota, a ranch with 20 mile sea beach. And Frank and Marge [Lloyd] to Santa Cruz. So now where are we. Probably working for the Post-Intelligencer. Does anyone ever hear at all from Jon anymore? Tellertons were in yesterday with that sweet Toby [dog]. In prospect each Friday for the past several weeks I've been headed for the Bay region—correct proof at Stanford and attend to other business at Bkly [Berkeley]—but btwn cats and rats and collecting and Ed Jr graduation and refinance mortgage here, I seem to get nowhere except too soon to the end of this card. Love

Dick Albee, the younger brother of writer George Albee—who maintained a close relationship with Steinbeck during the 1930s until Albee became jealous of his friend's success—moved to Monterey with his wife, Jan, in the 1930s (Steinbeck, *Life in Letters* 156). The couple were close friends of Ricketts until his death.

Paul Budd was married to Idell Henning, Carol Henning Steinbeck's sister. A biologist who worked for the California State Fish and Game Commission, he later moved to Hawaii, where he taught high school biology.

To Dick and Jan Albee
July 17, 1937

mailed Sat.

11:15 AM. Cannery Row humor: 2 workmen down in the street carting around heavy steel rails. One asked the other what time it was. Reply: "way after one o'clock." That's what I call delightfully preposterous, as though any 12-o'clock-quitting-workman would let that time go by. Man No. 1 nearly dropped his end of the girder to clout the other. I think the Greek letter pi, paragraph sign, would be a fine thing on a typewriter. In fact I'll have one put on my next portable, along with diacritical signs and some zoological symbols. It's "all quiet" until the glass case gets broken, either from outside or inside, and then maybe it's dynamite if the jedus inside happened to be sleeping or comatose instead of just an exhibit. I mean the dream.

Gret is back from the south, was in yesterday afternoon for awhile. Fred-Frances [Strong] house coming along at great rate. It will be [a] simple and beautiful place in the sense of beautifully simple. Saw Evelyn [Ott] who was in one of her periodical bad headaches. Saw Rudy at Shack. Pleasant time at your place, as usual. Long letter from Jn. Come along down. Love Ed.

Frances Ricketts, Ricketts's sister, married Fred Strong in 1932.

Evelyn Ott was a Jungian therapist and friend with whom Ricketts corresponded and socialized with in the 1930s and 1940s. He also saw her professionally. Ott practiced in Carmel.

To V. E. Bogard
September 19, 1937

Dear Bogard:
Will you enter our order for the following:

- 1 barrel formalin
- A quantity of chloroform, any efficient shipment up to 5 gallons— we use quite a bit, maybe 4 or 5 gallons a year
- 5 gallons glycerine

Terribly rushed; the boys are sending in frog shipments almost daily, over 700 yesterday. And cats and crayfish still coming in, but not more than enough for orders. This is the first time I have tried to hold it down alone during the busy season and it's just about ruining me. Would be bad enough anyway, but with specimens like frogs still coming in at this late date! I'm a wreck. But we'll be mighty glad for those frogs during winter and spring.

I hoped we could wait on the formalin until Oct. but I used the last today, and lord only knows what the expressman will bring in tomorrow morning in the way of frogs. Probably need more containers before the first, but I'll let it ride until I'm sure.

Ed.

Ricketts periodically employed local men to help with collecting. Often, these men were the "bums" Steinbeck would later describe in *Cannery Row*. Edward F. Ricketts Jr. confirms that "the boys" mentioned in this letter likely included Harold "Gabe" Bicknell, who appeared in Steinbeck's novel as Mack (Ricketts Interview, October 12, 2001).

The following letter is typical of much of Ricketts's professional correspondence; he often recommends other suppliers for those orders he is unable to fill.

To V. E. Bogard
November 28, 1937

Dear Bogard:
You can get all ready to crown me:

Your order for Santa Clara Univ. arrived in the midst of a welter of fresh clams, frogs and squid and I didn't have a chance even to look at it immediately. We have plain embalmed cats on hand and I could doubly inject two of them in about 3 or 4 hours time, but I can't find the time; there are still a lot of frogs to do. (By the way we have so many frogs now that we just sold a whole 50 gallon barrel of them to General Biological, wholesale). Also we have no Necturus whatsoever; we'll have some plain preserved specimens in soon—they are promised definitely. But they ask for doubly injected. Cannot supply—I'd have to order them in alive from Wisconsin and doubly inject them here and I just can't take the time now especially since we sell not very many doubly injected and it's a long difficult process. You might try the Bkly outfit for dbly inj embalmed cats, altho I've not yet heard a single good report of their embalmed cats. Doubt if they have any dbly inj Necturi. Shipment perch, Hydra expected any day, Grantia again on hand.

Ed.

The process of doubly injecting or embalming a specimen involves using a yellow fluid in the veins and red in the arteries. Edward F. Ricketts Jr. remembers his father embalmed sharks in the basement of the lab late into the night. "You can't believe how many nights I'd wake up from sleep and hear my dad downstairs injecting sharks. Cold with dim lights. God!" (Ricketts Interview, October 12, 2001).

In the following letters, Ricketts makes repeated mention of his feelings of overwork and exasperation about the financial troubles facing Pacific Biological Laboratories. At this point, as was the case throughout the nation, the Great Depression had impacted Ricketts's business, and he was faced with the possibility of having to close the lab for good.

To Jim Fitzgerald
February 21, 1938

Dear Jim:
I have been into a mess of overwork, trying to run this place alone even during the busy season, and have just been digging out thru the help of Gustav Lannestock who has been tearing into cleaning up my desk—finding among other things your most welcome letter written right after Christmas.

I see Evelyn only occasionally—Bruce is giving her son Peter lessons in boxing; Jean and I are again or maybe better say still, close and warm friends. If she comes down today later as I expect she may care to add a postscript. They are in their house up back of the Presidio; Bruce did a fine job of building. He is [a] good man.

Many things have happened since you left. Sue left Harry and is living in PG; painting on WPA project. Francis and Elaine [Whittaker] I see almost not at all; Jon and Alice Cohee not for a year. Jon fairly often; Carol is just back, I haven't been up there since. Hilary Belloc comes in occasionally. Red Black who with Polly is temporarily living in her family's Carmel house, say that Hilary went back to crab fishing. Frances and Fred Strong are living in their own recently built house in Carmel Woods, my mother with them. I see Nan and the children occasionally; they seem to be doing well.

My book still hangs fire; I corrected [the] galley proof on it now nearly a year back. Went down into lower California once last year, twice up north, the second time into Canada with Edw Jr and Nancy Jane. I have worked up more of the ideas all of us have discussed in essay form, may work them over with a view to eventual publishing. You might mention to Peggy that Shankar and his troupe of Hindu dancers and musicians gave a recital Saturday in Carmel; I was very moved. Expecting something superlatively fine, I was nevertheless agreeably surprised. I had furthermore a sense that I was touching merely the superficialities of a very deep field. Dick Albee has called up only Friday to say that this troupe was in SF, that he had press passes, and that I would do well to drive up there for it.

I have continued to add to my library altho I shouldn't have afforded a single purchase; picked up a new (to me, but actually old, Bombay 188 . . .) translation of the 12 principal Upanishads. An Oxford (Max Muller Ed) Bhagavad Gita. Best of all, a two volume English Plotinus. Also, most recent news, I have now a phonograph and radio receiver built by Pol Verbeck, who made Dick Albee's and John's fine sets. This is the best so far. Have some Gregorian and other old liturgical music, some quartets, Bach, Sibelius, etc.

Well I write at length. As I talk. Say hello to Peggy. I hope well for you, singly and both. I know you'll be having fun back there, this first winter for you now for a long intervention, in the cold and snow. The correlative here has been rain and wind, the Carmel pine trees went down in windrows. And we have been cold, in single board, cracked houses.

Ed.

James Fitzgerald (1899–1971) was a watercolor painter who achieved moderate recognition first in Boston and later in the Monterey area, where he lived with his wife, Peg (often called "Pegs"). In the 1940s he moved to Monhegan, Maine, where he continued to paint local landscapes and individuals in a unique style.

Gustaf Lannestock arrived in Monterey in the late 1930s and worked at the Del Monte Hotel as a masseur for a short time. He translated many of Carl Arthur Vilhelm Moberg's Swedish works, including his four-volume "immigrant novels," which included *Utvandrarna* (*The Emigrants*, 1949–59); *Invandrarna* (*Unto a Good Land*,

1952); *Nybyggarna* and *Sista brevet till Sverige* (translated together as *The Last Letter Home,* 1959). In his letters, Ricketts habitually misspells Gustaf as "Gustav."

Bruce Ariss (1911–94) was an artist who moved to Monterey with his wife, Jean, in 1934. The couple was likely introduced to Ricketts by Remo Scardigli that year. Bruce started a local newspaper, *The Beacon,* in which Steinbeck's story "The Snake" was first published. Ariss also started *What's Doing* in the 1940s, a magazine he edited and illustrated with his block prints.

Francis Whittaker (1906–99), a radical communist, lived and worked as a blacksmith in Carmel from 1927 to 1963. He then moved to Aspen, Colorado, and spent his life teaching as a blacksmith and traveling throughout the country.

Hilary Belloc, the son of British Roman Catholic apologist and poet Hilaire Belloc (1870–1953), lived in San Francisco and frequently visited Ricketts at the lab.

Uday Shan Kar was a Hindu dancer who toured the United States during the 1920s and 1930s. He was discovered by Solomon Hurok (1888–1974), who was also credited with bringing Mary Wigman (1886–1973), the German "mother of modern dance," to America in 1929.

Pol Verbeck was a Belgian resident of Monterey who periodically built and repaired electrical equipment, including the phonographs mentioned in the preceding letter.

⌒

To V. E. Bogard
February 26, 1938

Dear Bogard:
The fact of the matter is that I can't even touch the information needed for annual report still for awhile. I can't get enough sleep to be fresh enough for records. I must put up sharks while they are available, and I'm glad to say that some have come in. I was in an awful spot. Sharks, the fresh material of which costs usually under $200 per year, sell for up to $2000 per year, one year over $2500; and if I can't put aside some now for next summer-fall, PBL will simply be sunk. When I became fully conscious of why I couldn't

get them, and the seriousness of the situation, I went to bat. Got the Chinaman, simply thru friendship and good previous relations, to divert a limited quantity illegally from his contract with the shark liver oil cannery. This year I've thought it wise not to pay for any extra help at all, so I've alone and unaided, doubly injected and beautifully prepared every specimen that came in; terrible job. Working way into the night and then being routed out in the morning even before 8 by another load coming in. In the meantime, a good friend over in Crml, retired Swede [Gustaf Lannestock], very graciously has been coming over to the lab and going thru and filing my mail which I haven't adequately tended to for months. But he can't compile the necessary records. Everything's behindhand. So anyway, I don't want that good Chinaman to get into legal difficulties thru [a] gentleman's agreement with me, so I wrote the cannery, putting things up to them strongly but kindly, and they've been very cooperative. Would like me to be able to use specimens without livers, which I'd like to do also, because then a supply would be available almost at will, but I can't work out an adequate technique, and anyway schools would tear the air if they got specimens deprived of the most characteristic internal anatomy. So now finally the chamber of commerce; the president of whom is a warm personal friend of mine, Jon Steinbeck's brother in law is unofficially working on the problem, and they can bring so strong an influence to bear that I have some hopes there.

I might just as well face the fact that for several days still—even the last sales tax report not yet filed—I can't even start to work up the records; when I do I hope it'll go fast; then I can drive up there with the dope maybe. Stanford Press wants to see me too. It would be almost hopeless for me to try to make that report myself; maybe it'll have to go delinquent. Anyway I wanted to tell you about it so you'd understand I wasn't being delinquent entirely from carelessness. It's hard hard times. Maybe hope, maybe not; anyway best to go ahead and try rather than give up and swell the WPA ranks.

Ed.

The "Chinaman" Ricketts mentions was Chin Yip, a local man Ricketts periodically employed at the lab.

Steinbeck's brother-in-law was Bill Dekker.

To Dick Albee
March 4, 1938
Dick Albee

Well I guess I'm writing you because I called you by phone, or I intended to write you last week, or something like that. Anyway I'm writing and I don't know why I should worry why, or even how come I did. Or if I did. It's as Jon says "I believe him because I can't think of anything better to do and besides I'm too busy to be doubting things all the time." Just like people are not only mis-spelling firken all the time, they mispronounce it too. You know, a barrel, a hogshead, a rod consisting of 6 feet or is that a fathom. Fathom is good, good fathom.

Although I did go, I have really no need of going to hear Shan Kar as it turns out. Because I, my proud feather lord I can't spell that word feathers, have a machine [phonograph]. A deus ex. Pol done it. And he done himself proud, and me too. Now the neighbors know. They have suspected for a long time. In the first place, you can't tell them that I anaesthetize and embalm cats but to bring them back to life at will. As many times as I say "Now, when a cat's dead he's dead, and that's the end of him, no one can bring him to," they go away still shaking their head significantly. Pol hit the nail on the head; says I have a queer and secretive business with that-there-now science, that the seat of the business is located in a queer and out of the way place. That I am queer. And worst of all, I have books on magic, and medieval philosophy, and bibles, and I play Gregorian music and phonograph pieces with queer time patterns that make not only my windows rattle but Flora's as well. And if the gals over there can't rattle the windows in the pursuit of their no-doubt pleasant business—did I say pleasant or oh I was thinking of the peasant-pheasant confusion—no one can. But I do. So that proves it. I was thinking originally about my lovely new machine. And it's as good as it's beautiful. Well Dick to think I should find you sober.

There was another point to this letter, little as you'd think it. The Shan Kar dancing was every bit as fine as you indicated. In fact I went with the most very highest expectations, and still was agreeably surprised. If I had seen his troupe performing originally, 20 years ago, instead of opera ballet, I should have grown up with a better realization of the earnestness and significance of this form of expression. And would have been a better man, and more of me. I originally found it merely decorative and didn't know better until I saw Wigman and some of that type. He tops them. I had a very good seat right down front. Good thing, because some of the gestures aren't obvious. There is a Victor album now of Hindu music. When you come down here next I'll be the proud father of Stravinsky's Psalms. And not paid for. But charged to me, which eventually amounts to the same thing. I am also the father of a new essay. This one on poetry; good job I think. Some one ought to write an essay on poetry. I like it. As Jon's attitude towards sex "I'm in favor of it, I like it, it's alright." Belligerently. Now why should he be belligerent. Or was it the mayor who was belligerent. No, not to us. Bill Decca [Dekker], on the loose the other night, has hizzoner in. And we called up Jon. 1:30 AM. Oh wotta mob. Well Dick Albee and good Jan, I'm sleepy, now why should that be. Did I say really sleepy; well that's why the letter tails off.

> Flora Woods owned and operated the Lone Star Restaurant on Cannery Row, a brothel later depicted as the Bear Flag Restaurant in Steinbeck's *Cannery Row*.
> Bill Dekker, Steinbeck's brother-in-law, was married to his sister Mary.

To Dick Albee
March 29, 1938

I remember you very well Dick Albee; you lived in Jon's house from ... to ... I was also living somewhere then if I remember correctly; but I never do. If at all. Let's see, wasn't there a little woman too and some chick-a-biddies. Well I suppose all grown up now and gone to college; how time flies. I remember when your youngest, or have you had younger ones since, was christened. How she peed all over my shirt front. Or did I wear a shirt. The little bastard. Every time I show up at the cleaners, they say "what is that on your shirt, wine?" And of course I, not wanting to betray the ill breeding of your brats, say always "no" and let it go at that. But people wonder as you can well imagine. And it's got so now that I'm beginning to wonder myself. I always use the word pee instead of piss because it's so much more elegant. Everyone knows that only horses piss.

Anyway, if they show up here, we'll all get very involved and drunk; it will end up that I applied to SRA and gave you as a reference, but you weren't home when the investigators called, so we all got drunk again and were very very happy.

I'm sure from the way you were talking when I left, in a grey dawn, that you will have had a fine Chinese dinner. And no one knows how I enjoyed it. Also I note that the newspaper guild voted to strike; serves them right I say; if they want to work on a newspaper they should strike; I see the hand of Russia. But of course if the *Times-Picayune* tells you to strike, you just have to do it. And so do I, and so do I; some think the world is made for melancholy.... No it was just another wrong track, that damn switchman. It always tickles me to think of the happy happy people (this was in the old days of handswitching) who were carried on to the wrong street car line by a new motorman and conductor. I'll bet there was some sweating when the conductor finally called up the car barns and said "where am I?"

             Yours

  In the spring of 1938, Nan Ricketts moved to Vancouver, Washington, with Nancy Jane and Cornelia Ricketts. Nancy Ricketts remembers, "Dad drove us to the bus and looked teary-eyed when he said goodbye" (Letter, January 17, 2002). Edward F. Ricketts Jr. moved into the lab on Cannery Row in June of that year. He had been living with Fred and Frances Strong at their home in Carmel. A few months after Ed Jr. moved in, Ricketts wrote "Thoughts on My First (Substantial) Taking Over of the Responsibility of a Parent," in which he arrived at the following three conclusions about the goals of parenting:

1. To get the work done quickly, efficiently and pleasantly as possible
2. To have as good things to eat as our combined resources (or mine, originally) were able to turn out
3. To have as much fun as possible

Ed Jr. lived at the lab until he was drafted into the army in 1943.

To Howard Flanders
August 3, 1938

Dear Howard:
Many thanks to you for your thoughtfulness and kindness.

The embryo came thru in fine condition; I changed it over to 70% alcohol after several washings, and put it in a vial which is inside of another vial of alc[ohol] which in turn is inside a small jar of alc. I think nothing can likely go wrong with it until I can decide what best to do about it. It's about as small, maybe even smaller, than that 9.5mm specimens I got so excited about a few years ago, the smaller I ever got, the fine drawing of which unfortunately burned up.

I have had a great burst of energy since my teeth came out; don't know if it's due to the beneficial shock of the operation, to the removal of a veritable cesspool of absorbing toxins, or to a general waking up psychologically that I've been heading towards for a long time. Any way it's a good thing; I had plenty of places to put the energy. It's hard enough to keep going now even at best, and feeling dopey and half tired most of the time imposes a great added strain. Knocked off a great lot of proof on the index to my book and have been keeping up with the things here. I am going to the city tomorrow, let [the dentist] have a look at my jaw maybe pull out a few reticent processes, take some stuff to BKH [laboratory supply house] etc., and will stop in at Jon's on the way back. Will call up from there, and if Frances says you'll be down here for the week end, we'll get together. Hello to Emma and the family.

~

To Virginia Scardigli
September 13, 1938

Hello good Virginia: That was a welcome lot of stuff; I have been comparing the originals with this translation; very close, but of course anything [that] is changed often loses in translation. Many many thanks; I feel particularly grateful when I recall how much work it must have been, but I can quiet myself from being too much concerned by realizing I'd do as much and as gladly the other way around. Good for you, discovering "To Have and Have Not!" I had [a] similar experience at Jon's; sailed right thru it with renewed faith in Hemingway; I also had supposed he was slipping. The cheap concert business is one of the many advantages of living there; I wish I could get in on some of them, but it's so hard to get up there. I was there last week for a few hours to get me teeth; I had them put in then I thot, "now fine, a few hours free, I'll go see Virginia-Remo after I have a few minutes alone with this crockery so's to get used to it." Well that was more than a week ago, I'm still getting used to the damn things. I got so sick half hour after the installation that I was gagging and I went to a hotel and "be'd" properly sick for hours, but determined to keep them in as instructed unless it really go serious, and now the worst is over; they look all right, I can eat slowly, doing better (and more) every day, have learned to smile all over again, and sometimes now for as much as ten minutes at a time, well five minutes anyway, I even forget I have them. So alright. Cannery Row is going so strong it shakes the building. I am pretty busy. That Bungle is fine, shooting and stabbing a neighbor, so the other neighbors say, conspiring against him; his lawyer is worth the whole funnies. Dick and Jan were down last weekend. I haven't seen Jon for more than a month. The Bill-Alice et al combine was in a few days back. Brought me a

Beethoven opus 130 quartet in connection with having torn up the lab awhile back when I left them here pretty well drunk. I am on the wagon. Then nights in a bar-room playing; I think of you folks whenever I go by the First The —— [cut off]. I expect to be in SF probably just for the day, for a final tooth adjustment within ten days or so; this time I won't be sick, maybe we can go places. But whether or no, come along down as you are able to.

> Virginia Scardigli was a close friend of Ricketts from 1935 on. She graduated from the University of California, Berkeley, in 1933 with a degree in anthropology and in 1935 moved to Carmel. She held a number of odd jobs, including posing as an artists' model in Monterey. While modeling, Virginia met Fred Strong, Ricketts's brother-in-law, who introduced both her and her husband, Remo, to the lab group. Remo helped redesign the lab after the fire, adding the front bank of windows that face Cannery Row.
> Ricketts's mention of "comparing the originals with this translation" is unclear, as no earlier letter to Ricketts from Scardigli has been found.

To Torsten Gislen
December 27, 1938

Dear Gislen:
It was good to hear from you. I know how you will be having fun with the increased family at the holiday time. Christmas is essentially a family time, and I think it's even more significant in the Scandinavian countries than it is here. Funny thing, Ed Jr and I spent the afternoon and evening of the 25th at the home of some Swedish friends, a Gustav Lannestock who is living and writing in Carmel; we had goose and baked ham and imported pumpernickel and all the usual Swedish cheeses and meats and herrings and pickles. Before that we had spent Christmas eve at the home of Ritch and Natalia (don't recall if you knew them, part of the Kashevaroff outfit from Alaska), Xenia and John Cage drove down from Cornish School in Seattle, John Steinbeck and Carol came in and we

had a good reunion. Then when we came back from there very late, I switched on the radio idly and heard some (apparently) Swedish station broadcasting "the favorite Christmas Hymn of Sweden" some sort of celebration, with announcements in Swedish and English.

The business goes slowly, I have trouble picking it up since the fire, there is still insufficient money, the second depression is raging, I am getting along with out any help and have to do all everything myself. Which means that correspondence particularly suffers.

As unbelievable as it sounds, that shore book [*Between Pacific Tides*] will soon be out. It cannot possibly be held up much longer. I never saw such a slow outfit as Stanford University Press, and of course lately, since the fire, I've had to budget my time so carefully that it left little opportunity for me to do the things they wanted me to do in the checking etc. Now for six months it has been already in page proof, with all the illustrations in page proof also, and there only remained the revision and checking of the alphabetical index. I hadn't time for the considerable work required, but during these holidays friends have been helping me, and it will be done soon.

Nan and the girls are living in the Puget Sound country, I hear from them not often, and write infrequently. Ed Jr is living here with me, and it's a good thing for both of us. He's coming along fine, and I'm pleased that he shows no emotional marks from any troubles he may have been through. He has become one of the most considerate people I know, I have now a sense of being a very competent parent; he will turn out pretty well, well balanced and self reliant; promises to be a fine mathematician. I shouldn't be surprised if we hear from him sometime in the field of mathematical physics. Good thing!

I got a geoduck for you, and will sometime devise an effective method of shipping it. I have had neither the morale, nor time, nor money to start rebuilding my scientific library, but in the fields of poetry and philosophy it has crept up to pre-fire status. Collecting has been going well, but many things haven't been replaced and I fear never will be; I made a trip down south, covered pretty well the ground we went over on that fine trip. I have now a Ford 60, the small V-8, the cheapest car to run I have ever used. So I went from the largest to the smallest!

Now it is warm and sunny; the canneries are going strong—they will extract every single sardine out of the ocean if legislation doesn't restrain them, already the signs of depletion are serious. Funny how Americans can't learn the lesson that the north European countries have known for a century.

Say hello to Mrs. G. for me. I am remembering you folks very happily.

> Ritch Lovejoy met Natalia ("Tal") Kashevaroff when he went to Carmel to visit his friend and former teacher Jack Calvin, who was a close friend and collaborator of Ricketts. Calvin was married to Tal's sister Sasha, another of the six daughters of Father Andrew P. Kashevaroff (1863–1940), a Russian Orthodox priest. Ricketts referred to them as the "Kashevaroff outfit."
> Xenia (another Kashevaroff sister) and John Cage (1912–92) married in 1935 and divorced ten years later. In 1938 they were living in Seattle, where he taught at the Cornish School. Cage, a renowned composer, writer, and lecturer, achieved national recognition and honors, including a Guggenheim Fellowship, an award from the National Academy of Arts and Letters, and a Notable Achievement award from Brandeis University.

1939—1940

To Howard Flanders
February 15, 1939

Dear Howard:
I feel very grateful for the embryo you sent me. If I can get Ritchie integrated to do it, I want to have him make a drawing of the whole thing intact, chorion and all; and then another one of the embryo dissected out in the membranes. Little enough work seems to have been done, except of the most qualitative sort, in this field important to human welfare; I wish I were equipped, mentally and otherwise, and had the time to do it.

Have had the usual busy time. Finally, with the help of a couple of Seattle friends down here for the holidays, we finished the third and last revision and complete rewriting of the alphabetical index of my book. Now even that is in dummy, all proof read. Now all that Stanford University Press has still to do, so far as I can see, is to print and bind. Good job well done. I keep looking forward to the time when I can get a few days off for chasing around and visiting, as in the old days, but sharks come in, and work on the books has to be done for income tax, and increasingly I have to do all things alone, so it seems there won't be any spare time soon. Frances says it's close to Emma's time and I suppose you'll have a sense of anxiety, or at least waiting. John gave me a new Bach Concerto and there is some Russian liturgical music; come hear it when you get a chance. Ed is developing into a great (just that if persistence is a factor!) player on the trumpet which some not too-good friend of mine gave him.

The following letter reveals the kind of friendly frustration Ricketts felt when dealing with Gabe, later depicted between Mack and Doc in Steinbeck's *Cannery Row*.

To Harold Bicknell
February 18, 1939

Dear Gabe:
Persistence so excessive certainly ought to be rewarded. You certainly have been wanting terribly to collect Styela despite 5 or 6 statements from me that I'd already tied up enough money in them. Well, I seem to have tied up still another $1.07, 57¢ for express and 50¢ for a telegram. I have accepted them and will allow you something for them; there were 227 of which the majority seem to be good sized.

DO NOT UNDER ANY CIRCUMSTANCES SEND IN ANY MORE STYELA UNLESS I ADVISE THAT THEY ARE NEEDED.

The only things I want from down there that you can get and that can be shipped successfully are: Ischnochiton and brittle stars. Both are exceedingly common wherever there is suitable rocky beach with boulders partially buried in the substratum.

As I said before: I cannot advance any more money. If I wanted to today I couldn't anyway. I have an overdrawn bank account. If you are still in that area and really are willing to go out and get these two items, drop me a line a day or so before the next set of tides, and I'll send you a few dollars if I can spare them.

You won't be able to get anything on dragboats (that will stand shipping up here, although lord only knows there's lots of that stuff that I need) except what the sailors call "sea mice," an annelid worm (little as it looks like it) 5 or 6 inches long and about half as wide, upper surface covered with bristles usually full of mud; about the shape and size of a mouse, and gray in color. I can take some of these, 100 or so at 5¢ each delivered here. Any brittle stars that you get up from deep water will probably break up before or during preparing. I want sea fans, sea pens, sea "mushrooms" etc, lots of things that may be taken by dragboats, basket starfish, etc, but none of them will stand shipment, and there's no point to your going out from my standpoint. If I had the time and the money to go down there, you might bring back some stuff to me on shore quite profitably, but it will not stand shipment, and all that'll happen will be that the express company will make money on the shipment. If I had time and money, I could get the local dragboat men to bring things in. Later on in the year I'd like to drive down there to pick up some of the southern dragboat stuff, some of which may differ from what we get here, but I can't do so now.

If you go out on such a trip, it will be for your own fun. The only thing I can use that I know will probably stand the trip up here is the sea mouse, and I doubt if you'd find more than a few dozen of them if any.

After today (Saturday) and tomorrow, there won't be any tides even remotely suited to getting chitons for another ten days. If you're still around there then, and can get transportation to collecting grounds, let me know.

             Sincerely,

To Virginia Scardigli
April 25, 1939

Dear Virginia:
What is tides, and why are they between Pacific? As for your friend Miss Oodle-oop, you tell her for me that there are things far more pleasant, yes and I mean it too tell her and you and you and you, than bites in false rubber impression wax; and far better ways than that, yes far far better, of getting to New York, or new anything.

Yes I heard there was a book out. And little as you'd believe it, I was one of the last to know. With its customary delicacy, Stanford kept its midwifery hidden from me, until Dr. Light at UC who is already using it in his classes, wrote me a letter of congratulation. But then, when I wrote to Mr. Stanford himself suggesting the mother-father would like to see the bouncing child, they advised that Light had received [an] unbound copy only, so it's alright.

Heard about you good people pleasantly thru the Pauls and later thru John. Now for a long time I have just sat here without leaving PG's confines. Someone told me gasoline is expensive and I've been afraid to try it. May get down south collecting next month, tho. Once recently up to Los Gatos. There is half a plan afoot for me to go up with what's her name, Mrs. Gregg's daughter, to see [*Of*] *Mice and Men* [1939] later on with some friends of hers, but if I do it will be one of those up and back things and probably won't get to see you. Next month tho I hope—and how often I hope those things!—to be able to drive up there unhurriedly Fri. afternoon for return Sunday; maybe get my teeth relined (dumb I never did get around to doing that! Scheduled for Jan. or Febr.) and just sort of visit around. Oh, yes, did you notice, the line drawings in the book were by Ritchie, and weren't they good! Nice if I'd make some money on the things, but no royalties at all for first 500 copies, then 10%. So if, in my lifetime, I get back 10¢ an hour for the time put into that job, I'll be lucky. But even that's better than nothing. Anyway, come down when you can; we'll ship Ed Jr over to his aunt's house in Crml, and you folks can sleep in the same bed that was once at your place; so that again, all the noisy night, you can look across the street to see what goes on in the whorehouse, and maybe we'll get Bill Dekker to call up and Remo can talk Italian to him and we'll all be happy.

> The "Mrs. Gregg" Ricketts mentions was Mrs. Hattie Gragg, a Monterey woman who, along with Susan Gregory, helped inspire Steinbeck's *Tortilla Flat* (1935) with their stories about the *paisanos*. Mrs. Gragg's daughter was Julia Breinig.
> *Between Pacific Tides* was published in mid 1939.

To John Steinbeck
July 2, 1939

Dear John:
Stock certificate enclosed. There was some delay, and there may still be a little mix up tho I don't anticipate any. APF [Austin Flanders] lost some of his stock, [I] think it's all straightened out by request for affidavit of loss, etc.

When Bogard and Frances and I sometime are all in the same place, we'll have a meeting to elect you vice-president and director to replace APF, who automatically loses that status by selling all his stock. However no hurry.

I have been trying to work out some way whereby the thing will stand on its own feet and you will be protected if you inject some new cash into the firm.

Your taking over the about $4000 mortgage will be a businesslike procedure; will stand on its own feet, be a good thing for both of us, since the security is good, since you'll be getting no less interest. And the security surely must be sufficient or the bank wouldn't lend the money on it.

But neither the taking over of the mortgage, nor your acquiring Austin's interest will provide any added capital. Maybe even doing that won't do any good—altho I think there's a good chance—but there's certainly no point to hoping anything from the firm unless some capital is available. The amount you originally discussed is, I think, needless, it couldn't earn its share even of interest let alone dividends. But some extra money is justified from a business standpoint.

Well, I thought the whole thing might work out mutually well, and protect you, if you were to advance $6,000 instead of $4,000 on this mortgage. That extra $2000 might do the work. And at 3% that would be $180 per year, or 15 per month interest. We're already paying that much or more, so there'd be no hardship there, with the added advantage of something to work with. But at the same time you'd have some, even considerable protection, in the case of my death or if the lab went broke. And some sure income from your investment. I don't suppose we ought to figure on starting to pay back any of the principal at first, but so long as the interest was paid up, the debt wouldn't get any greater.

That way, at the worst, you'd stand to lose only your stock, plus the difference between what the mortgage is actually worth, somewhere between $4000 and $5000, and what you will have loaned on it. If I die, or if the firm goes haywire the stock may or may not be valueless; I don't know.

Also I don't know when, if ever, the firm will again be in a position to pay dividends on stock. Would be only a question of building it up (don't know if that can be done, but it doesn't look hopeless), to the point where it can pay me $250 or so per month salary, that's really all I need except for some special needs such as [a] library for further investigations, and maybe my accrued back salary. After that, if it does go ahead, balance can be used for dividends; or some of them (as in the case before the fire, when they burned up) for additions to equipment and building.

> In early 1939 Austin Flanders sold his Pacific Biological Laboratories stock. At Ricketts's request, Steinbeck agreed to pay the balance on a loan Ricketts had taken and become the lab's vice president. Steinbeck remained a silent partner in the lab until Ricketts's death in 1948.

To John Steinbeck
July 13, 1939

Dear Jon:
I saw Toby [Street], and he yesterday started things moving on the mortgage arrangement.

However, it'll have to be the long way around—which I hoped could be obviated, chiefly because it will cost PBL quite a bit in legal fees, title guarantees, filing fees etc. But he says there's no other way to make it legal and to provide protection for you in case I should die, etc.

I suggested a mortgage based on $6000 at 3%, interest payable monthly, which will amount to $15.00 per month, and no repayments of the principal for 2 years, and then at the rate of $25.00 per month, which means $40 per month payments starting Sept. 1941. That's the legal requirement. However if by then it still seems unwise for us to divert any principal funds, we can consider another year or so just keeping up with the interest, and no legal changes will be required if you don't complain legally.

Now in order to do this, a stockholders meeting and then a director's meeting will be required first. And since 10 days notice is required by our by-laws, nothing can happen until towards the end of the month. Then Toby will handle the whole thing at the meeting, put an escrow right thru, and will place it on record as completed July 31st. The stockholders meeting is required because we must have a quorum of directors, electing you to the vice-presidency and directorship left vacant by Austin Flanders, to pass a resolution permitting me to effect this mortgage. Then Toby says that the only way is to have new title guarantee. That's unfortunate, suspect it represents a (maybe necessary) racket on the part of the title guarantee companies, since every time property changes hands they get a new crack at it. In the last ten years we have paid for three or four searches of title on the same property, just because we sold some of it, or established a new mortgage. But he says there's no other legal way out of it and he surely knows. Seems to be a much more careful and meticulous person than I should have suspected.

Maybe you'll be down here today or tomorrow, but if you don't, why this is dope to date. I hate delays, like things simple and quick in business relations, but where mortgages and corporations are concerned, it can't be done without a lot of red tape. However this way you'll be protected in case of my death, and 4 or 5/6th of your mortgage investment will be entirely safe and sound. So, see you soon.

To Joseph Campbell
August 25, 1939

Dear Joe:
I shall be perfectly delighted to see you.

A good many things have happened to me also in these seven years. Externally most important: little less than three years ago the lab burnt down [as a] result of electrical fire in cannery next door which swept this part of the waterfront. Everything I had [was] destroyed, library especially. Just before that Father died, whom you will remember as my right hand man. And a little before that Nan and I had a permanent split-up. Ed Jr is with me, getting along fine; the girls are with Nan in Washington. My book finally went out, as you no doubt will have seen. When my files burnt up in this holocaust, I lost your address and so lost possibility of touch with you. Ed Jr only was here when your friend called up and I never did get even his name.

In continuation of the things we discussed, I have worked up three essays that pretty well sum up the world outlook, or rather the in-look, that I have found developing in myself more and more during the years. Unfortunately for the progress of things of that sort, however, I have had such an adjustment problem to finances and personal difficulties that much of the time I haven't either energy or morale for acuteness of this type. I have another sociological book projected; big job, mostly hack work, don't know if I'll do it or not. Loss of my library was a blow. I got another Spengler, several Jung things, some medieval philosophy, but mostly poetry. I got my morale up to the point of working on new scientific library only recently.

Xenia and John Cage, whom she married—modern composer, are in town. Ritch and Tal back here to stay from Alaska. He is becoming [a] consequential poet. John and Crl just left; will be back for weekend. Most of the news. I'll be so glad to see you. And I have a sense of welcome towards your wife.

In February 1932 Joseph Campbell (1904–89) moved into Canary Cottage, the house next door to Ricketts's own house on Fourth Street in Pacific Grove. At that time, Ricketts was still living with his wife, Nan, and their three children, although they experienced periodic separations. Ricketts and Campbell became close friends, socializing with the Steinbecks, Lovejoys, and other members of the lab group.

According to excerpts from his journal, Campbell and Carol Steinbeck fell in love during the spring of 1932; however, both agreed that ultimately friendship was the only relationship they were meant to share and ended the relationship. Campbell had accepted a job teaching at the Canterbury School in Connecticut, beginning in September of that year.

In June 1932 Campbell joined Ricketts, along with Sasha and Jack Calvin, on a collecting trip to Alaska. The four sailed Calvin's boat, the *Grampus,* from Puget Sound to Ketchikan, Alaska, spending more than two months collecting both for Pacific Biological Laboratories and Ricketts's own personal research. During this trip, Ricketts discussed with Campbell his concept of nonteleological thinking, or "is thinking," a dialogue that continued in their correspondence after Campbell's departure from Pacific Grove.

Campbell married Jean Erdman in 1939, and the two traveled to Hawaii to visit her sister that summer. On their way back to New York, they visited Ricketts. It was their first meeting since 1932.

∽

During their visit, Ricketts gave copies of his three essays—"The Philosophy of 'Breaking Through,'" "The Spiritual Morphology of Poetry," and "Non-teleological Thinking"—to Campbell. In his letter to Ricketts dated September 14, 1939, Campbell remarks on the essays. "I am going to let your three papers serve as excuse for three more letters. [. . .] I am even a little glad, now that seven years went by with hardly an exchange of words between us; for, if we had been discussing our problems as we encountered them, nothing quite as pretty as this could have taken place! [. . .] Jung's 'Secret of the Golden Flower' will arrive in a week or two, from Jean and me, in recognition of a superior evening, night, and morning, and in anticipation of renewed fellowship."

To Joseph Campbell
October 7, 1939

Dear Joe:
Your most welcome letter came some time back, and the book arrived this morning. Many, many thanks to you. I had been dallying with the idea of getting the *Secret of the Golden Flower* [Jung 1938] for a couple of years; it's hard to pick up such out-of-the-way things locally.

I have started attempts toward publishing those three essays. Sent them first to *Harper's* and they came back just a couple of days ago, and only a few days after Jon who was with de Kruif in Michigan, had wired me for dope on the one on non-teleology and had then, with de Kruif, unbeknownst to me, wired to *Harper's* asking them to make a typed transcription of the essay charged and sent to Jon. He apparently thot that that influence would call the things to their attention, in addition to his wanting a copy of the revised essay.

Now I try *Atlantic Monthly*, then *Yale Review*, then *Southern Review*, then *Partisan Review*, but I think none will take them. So there I'll be. Guess I'll have to be well known before I can publish the darn things, and that may never be.

I am to do another job for Stanford—a manual of the invertebrates of the SF Bay area. Won't take long, a year or so if I can get to it. And there are a lot of ideas coming up based on those indicated in the essays, but that's slow and spotty. Tal is having more fun even than before with *Finnegans Wake;* I have been reading the new Wickes "Inner World of Man." Bureau of Fisheries assistant from Manila has just been here, leaving me more than half tempted to go over there for a couple of months collecting etc when I can raise the steamship fare. I had planned to spend next summer up north, but now I think I may stay here after all. So come along and spend the vacation in PG's good dripping fog. I liked your Jean very much as you both no doubt realized. We'll take lots of pleasant trips; I know the mountains hereabouts now fairly well.

To Xenia Cage
October 9, 1939

Dear Xen:
I'm serious about this Yavlensky business. I can't get that one head (the pastel colored one) out of my head; or rather I can't get the idea out, or the idea of the idea, something as involved as that, since I've lost memory of the appearance of [the] picture itself.

If you'll send me that woman's name and address, I'll try to plan things so as to stop in there. May be out of sight for me to buy one of those, but anyway I can consider it. What do you suppose she could let me have one of them for, in case I find one that clicks so deeply with me as yours did? Not your head. Well, that too.

What reminded me of all this I guess, was that I just stopped for a bite, and got to playing Russian liturgy. I have several new records—all that are listed to date I imagine, and that's few enough.

I haven't seen Jon since you did, but received a couple of screwball messages from him, so I guess everything's alright. Stanford Press has a new job coming up for me, if ever I can get integrated to do it; a short manual of the important SF Bay area marine invertebrates. My three completed essays, now much revised and polished, are out beating around trying, probably unsuccessfully, to get published. Joe Campbell and his wife were on here—did Tal tell you?

Very pleasant visit. He sent me Jung's *Secret of the Golden Flower* when he got back.

Well that's the way things are, you see, Xen. Just hammering in the lab, and tootings on the trumpet, and the building shaking from the machinery next door (the canneries started up as soon as war was declared, the first of the profiteers), and drunks chasing each other up and down the street, and Flora's gals "going and coming without hindrance . . . and the vermilioned girls getting drunk about sunset" and other times too. I am a little bit off the wagon—just a leetle bit. So I'll see you Christmas—not so sure now about coming up there next summer. But possibly, still.

> Aleksei Yavlensky (1864–1941) was a Russian figurative painter influenced greatly by Matisse and Kandinsky. In her letter responding to Ricketts, dated October 13, 1939, Cage references the art collection of Mrs. Scheyer who lived in Los Angeles and was likely the woman Ricketts refers to in the second paragraph of the preceding letter.

To Division of Fish and Game
San Francisco, California
October 11, 1939

E. L. Macaulay, Chief of Patrol
Division of Fish and Game,
Ferry Building, San Francisco, Calif.

Dear Mr. Macaulay:
I have awaited answering your letter of August 23rd until I should have assembled all the facts in the situation.

If many more frogs than the permitted thousand have been taken, it has been without my knowledge, and they haven't been paid for. Mr. Oxford is the only person to whom I delegated frog collecting this year, Ray Meredith having left permanently for Oregon early in the spring. Oxford's frog remittance summary is being attached.

However, I find that 1022 specimens were so checked in this year, 22 more than our permit called for. I failed to realize the total was so high for several reasons: because it was mixed with credits for crayfish, lamprey larvae, fresh water mussels and a variety of marine items; because, with two exceptions, the lots of frogs were small; and because in past years we have never thus succeeded in procuring anywhere near the desired amounts. However I am certain that your representative is incorrect or misinformed in stating that considerably more than a thousand have been taken. Some may have been eaten, and any that died en route or during storage were discarded, but that number cannot have been great. I suppose there's also a chance that specimens were sold elsewhere which should have come here, but I doubt it. Oxford is a pretty honest fellow for a cannery work type; he says definitely not. And I furthermore don't know where he'd sell them except to some of the wholesale frog dealers who could pay probably not much more than I do.

Incidentally, on the last trip to the valley, on which I haven't the exact dates, the boys got only 38 frogs—Cleo Oxford, his wife, and probably one other boy, having been gone for several days. They report no frogs available then. And I think it was on that trip that they reported that one of the dollar a year men had been quite upset—presumably the one who previously arrested them for not having their permit along (I gather he had been drinking, and in his cups may have been inclined to be authoritative); said he was sure they were taking more than their permitted number. If so, they certainly failed to arrive with them.

If there were some way whereby specimens could be checked in over there, difficulties of this sort could be solved, but much of the work is done inconveniently at night. In connection with any future possibilities, I shall be glad to have your reaction to the concepts embodied in the attached letter also.

Sincerely,
E. F. Ricketts

On October 14, 1939, E. L. Macaulay, chief of patrol of California's Division of Fish and Game wrote to Ricketts the following: "I am interested in the report that one of our so-called dollar a year men had been drinking when he ran across your collector. Could you advise me the name or badge number of this individual?"

To Division of Fish and Game
San Francisco, California
October 17, 1939

E. L. Macaulay, Chief of Patrol,
Division of Fish and Game,
Ferry Building, San Francisco, Calif.

Dear Mr. Macaulay:
I discussed with our collector the matter of the deputy, who, allegedly drinking, had been concerned with the number of frogs we were taking. Oxford says that this man turned out to be a pretty decent fellow after having had proven to him that our men weren't intent on breaking the law. His name is Charlie Viera, of Atwater; he drives a Signal Oil Company truck. On the occasion referred to, he was apparently pretty well inebriated, the boys say he fell into the river and had to be hauled out. However I can understand that on a cold night, an outdoor man might go over the line a little in the matter of drinking; I shouldn't tend personally to be upset by that. I am concerned, though, with any attitude of ill will that might arise on the part of the sportsmen over there as a result of our taking frogs. Viera told our man Oxford that the sportsmen had planted those frogs over in the San Joaquin, and that there was an element of unfairness in our taking them. I supposed that, although all the Rana catesbiana in California have been planted originally, they had spread and had now become indigenous. I don't want the boys to get PBL and themselves into bad grace over there. But if, as I had supposed, these frogs have acclimated over there, and have become legally and morally no one's property, it would seem that if the Division of Fish and Game finds it in order, we could go on tak-

ing them. There will be no activity now probably for the several winter months, but if starting next year, you see fit to renew our permit, I'll try to get Oxford et al to pick up two or three thousand specimens.

> Appreciating very much your
> cooperation, I am,
> Sincerely,
> E. F. Ricketts

On October 20, 1939, Paul de Kruif wrote to Ricketts the following: "John Steinbeck has got me all steamed up about your philosophy as a biologist—if that is not a contradiction in terms! I tried to get your 'Non-teleological Thinking' from *Harper's* mag. Editor Lee Hartman tells me that, under title 'The Philosophy of "Breaking Through"' it has been rejected. I wonder if you'd let me see a copy—there's just a chance I might land it with another mag for you. And I'm keen to read it before the great pleasure of meeting you—which our mutual friend John promises me."

To Paul De Kruif
October 28, 1939

Dear Dr. de Kruif,
By a coincidence, I was at John's place when your wire arrived. The essays in question had just been returned by *Atlantic Monthly*—apparently unconsidered or maybe even unread there also as the evidence indicates was the case with *Harper's*—so I just left them there at Los Gatos. John wanted in any case to read them over in their present maybe-definitive revisions. He was to send them to you direct. And since he tends to be both prompt and reliable (qualities I envy in him and intend always to start emulating!) you may have them already.

With reference to the possible relation between philosophy and biology, wouldn't it be a great thing if the theoretical economists, the sociologists—even the social workers and case investigators, arm-chair theorists in ethics and philosophy, would *have* to pass a good stiff course in field zoology! I used to think of that when a dear old lady used to come here investigating for county charities in connection with a fine down-and-outer-drunk who used to work here. If she could have known that most larvae perish in the hard struggle, that many failures is the price of one success, she wouldn't have been so lost whenever her pet rehabilitation-project fell off the wagon. But I suppose you can fill a cup only as full as its capacity, and lots of things she could never know.

John has mentioned you so often that I'm looking forward to your next trip out here; as I conceive the weather around Holland, Muskegon, etc, winter would be a fine time to be away.

I am very thankful to you for being interested; I hope only that John hasn't "blown up" these ideas in anticipation so as to leave the reality too pale!

           Sincerely,

Paul de Kruif (1890–1971) was a bacteriologist who taught at the University of Michigan from 1912 to 1917. He wrote numerous books, including *Microbe Hunters* (1926), *The Fight for Life* (1938), and *Hunger Fighters* (1939). Perhaps his most significant contribution to literature was influencing Sinclair Lewis and assisting him in writing *Arrowsmith* (1924) by providing technical and scientific details. De Kruif and Steinbeck met in 1939 while working on the film version of de Kruif's *The Fight for Life,* and the two remained friends thereafter. According to Edward F. Ricketts Jr., "At one time [it was] being bandied about that Paul might accompany EFR and JES on the Sea of Cortez trip" (Ricketts E-mail, August 11, 2000). He did not go, however.

Ricketts was fascinated with music and was known to have compiled a large collection of phonograph albums, particularly Gregorian chants and classical pieces such as Bach's *Art of the Fugue.*

To The Gramophone Shop
New York, New York
December 9, 1939

The Gramophone Shop,
16 E 48th St.,
New York City, N.Y.

Gentlemen:
I am anxious to pick up the two records mentioned below, one of which at least is out of print. It seemed to me that you might have them or be able to pick them up.

The one, which quite probably you have in stock or can readily pick up is

Columbia No. 68084, Bach Brandenburg No 3.

Unfortunately I haven't the number of the other. It's an Okeh of Russian liturgical music, with Ochi Nash (our father) on one side, Blasen Mux on the reverse. If you can spot the item by this description—unaccompanied chorus—I shall be glad to have a quotation on it.

Sincerely,
E. F. Ricketts

⁌

In his letter of December 10, 1939, to Ricketts, Campbell comments on "Non-teleological Thinking": "Page 13—do I recognize myself in the very intellectual youth who had led a cloistered life and determined that what I needed was to suffer?—(Bull's eye!)." The passage in "Non-teleological Thinking" to which Campbell refers to reads: "Suppose a very intellectual but unbalanced youth, who had led a cloistered life, were to determine by scrupulous observation that maturity and charactral integration followed only in the path of pain and struggle, were to say 'Well, what I need is pain,' and were to put his hand deliberately into a flame. The unlikely maturity achieved that way would be ridiculously round-about, might easily defeat its own purpose" (Hedgpeth, *Shores* 2: 76).

To Joseph Campbell
December 16, 1939

Dear Joe:
I am thoroly thankful to you for the suggestions regarding the first essay embodied in your letter. I agree entirely with those I have gone over; for the rest I am filing your letter with the essay itself. There is still a more recent edition than the one you have. Paul de Kruif has it now, and seems to be giving it, with the two others, very careful consideration. The consensus of opinion seems to substantiate what you also say, that publication in a book offers the best chance—along with a couple of other essays now in the offing (one on the relation between the user or the taker and the producer or giver, as illustrated by the distribution of electrical power; understandings of the relation between the giver and the taker are common, but between the taker and the giver not common). And for this, I'll probably have to wait until I become known, if ever. *transition* [literary magazine] however does offer a possibility, altho their trend is toward new form, rather than considered content.

I hope, not entirely unselfishly—since I'll be very interested in the results—that you come out successfully in your writing job. Those ideas ought to be worked out verbally and recorded so as to be more generally available; seems like a Herculean and to some degree a thankless task, all of which it is; but it should be done nevertheless. I envy you your facileness. The answers certainly seem to be in some of the Oriental philosophy. I have two new translations of the *Tao Te King;* the one in the *Bible of the World* seems to me an easy and interesting introduction to the ideas; other more profound translations being available for later consultation.

No Joe, I hadn't thought of you at all in connection with the p. 13 reference; I don't even know if it's apropos or not. Peoples' paces differ; mine seems to be slower than the average, or more intense, or something! I have had more suffering in the past year than in all my life up to then. But there's a sense of construction as well, and I suppose everything will have been fine if it ends fine. And it's something to have one's life improve in a financial sense and in methods of intellectual expression, which is where I seem to be improving.

Say hello to Jean, write as you find time. And thank you again for the very apropos suggestions. Tannhauser has just been coming over the radio, and I have been thinking of the words "the hills of home"—don't know why.

> Campbell was working on a critical analysis of James Joyce's *Finnegans Wake* in collaboration with Henry Morton Robinson. The volume, titled *The Skeleton Key*, was published in 1944 to much critical acclaim.

To Pascal Covici
December 23, 1939

Dear Pat:
Many thanks to you for that good book [*Caribbean Pleasure*, 1939]. I read it through immediately and thoroughly, a thing I seldom do. Good job.

I had a very pleasant letter from Mrs. Covici. Was finally able to dig up a little more of the [shark liver] oil, and have suggested that she continue giving it to her mother, despite lack of results thus far.

John and I start out today on some preliminary work—collecting here in the SF Bay area, but since the rain is coming down in a steady stream, I don't look forward to it very happily. However the Mexican trip promises well from that standpoint; after a few days of that hot sun, we'll welcome any sort of change. Probably, as on most difficult trips, we'll have lots of fun and lots of grief.

We had a good time when you were here. Come again. Next time we'll get together and drink a little wine!
                                Sincerely,

Pascal Covici (1888–1964) was Steinbeck's publisher and close friend from the mid 1930s until his death.

To Dorothy Covici
December 23, 1939

Dear Mrs. Covici:
Since the air mail package, which was so unfortunately damaged en route, was sent to you, I have been on the look-out for more oil; a couple of days ago, coincident with the arrival of your very welcome letter, I was successful.

Another bottle of this is being forwarded. Altho there may not even yet have been any appreciable results, I would suggest that your mother continue with this until the new bottle has been taken. Then, if it seems to have done any good, I can forward more. If it hasn't no harm will have been done. I have come to look upon this oil—which is from the liver of the basking shark which feeds upon microscopic floating plants (one of the largest sharks known, up to 20 feet long, yet its food is so small as to be invisible!)—as an almost magic relief in arthritis and in some types of allergic disturbances. If it has no effect in this case, this will be its first signal failure. We can hope it will work.

Today I am writing also to Pat, who forwarded their recent publication *Caribbean Treasure* [Ivan Sanderson, 1939], a fine job. All of us had a good time when he was in the Grove a while back; I hope he can repeat the trip.
                                Sincerely,

To Peg Fitzgerald
February 6, 1940

Dear Peg:
I am grateful to you for remembering me at Christmas, and sorry that my procrastinating proclivities have let me hold off answering so long.

I see most of the old bunch pretty often, Hilary and Hope [Belloc], Frank and Marge [Lloyd], John and Carol—went to pre-Cascarone Ball the other night at Greggs—Jimmy and Hattie led the grand march. Ritch and Tal came back from Alaska. There are a few new people. Sometimes we have lots of fun. I see Jim off and on, sometimes not for months, sometimes once or twice a week. Hard to tell if he's happy or not. Jim is a curiously aloof man. Sometimes I think he doesn't deeply know about such external realities as events and people. But he seems to be having fun, and, altho I haven't been around to the house almost for years, I understand he's been doing some significant painting.

Lots of rain this winter. Which means that next month the hills will be green. And the wildflowers [in] back of Highlands and on Big Mountain will be fine.

Jon and I are working up the first of two books together. It goes slowly but what we've done so far I think is good. Planning a very extensive trip into the Gulf of California for next month; 6 weeks or two months.

I hope you will come here. I'll be glad to see you.

> Peg and James Fitzgerald were married in 1931 in Del Monte, California. In the mid 1930s, the couple separated, and Peg went to the East Coast. They attempted to reconcile but never maintained a successful marriage, permanently dissolving it in 1944.

To Herb and Rosa Kline
March 7, 1940

Dear Herb and Rosa:
We just read your very welcome letter, and Jn asked me to reply. Carol is back at the ranch packing up for the trip; Tal went along. Jon and I are putting a few last minute touches on preparations here.

First: our plans had to be violently changed still once more. The boat we had lined up had a change of heart at just the last minute, thought a few dollars more per day would be just lovely, especially since we were in such a hurry and therefore couldn't look around to get another—typical Sicilian stunt. So, we did get another, a far better boat. Charter all signed. Thank dee lawd for the Jugoslav fishermen. Their initial moroseness and apparent suspiciousness followed by cooperation and genialness is a pleasant contrast to the blatant friendliness of the Sicilian which doesn't hold up.

So the call letters we gave you are no good. The new boat has no San Pedro crystal—operated in Alaska waters with Seattle crystal, so no long distance ship to shore telephone conversations are possible. The new boat is the *Western Flyer*, call letters WPTO—but good for short wave radio only, not for Bell Telephone System.

We can be reached via letter or telegram within a few days at La Paz, Baja California, Mexico, from March 19th to about the end of March, then on at Guaymas, Sonora, Mexico. At Guaymas we'll get in touch with American Consul first and last thing.

That party wasn't really over until yesterday. People kept dropping in, the merry merry drinking went on. In fact I'm not yet sure it's over. But we have managed to get the necessary work done, I wonder how and when. The answer is that happy happy people work rapidly, efficiently and integratedly—no conflict outer or inner. I have a pretty strong notion that when we show up on the waterfront at San Diego next Wednesday or maybe Tuesday evening, there'll be a partying delegation from Hollywood. But if we can [find] someone in the gang orderly enough to count and discriminate so that we leave San Diego, not only with the same *number* of people who arrived, but with the *same* ones, I guess we'll get thru it. And if this sober letter doesn't prove something, I don't know what. Nice people, I hope we'll see you soon.

Ed & Jon

On March 11, 1940, Ricketts, Steinbeck, and six others left Monterey on a collecting expedition bound for the Sea of Cortez. The group spent close to six weeks collecting and exploring the region. Ricketts's notes were later incorporated into the *Sea of Cortez* (1941) along with his essay "Non-teleological Thinking" in its entirety as the Easter Sunday sermon in chapter 14.

Shortly after their return to Monterey in May, Steinbeck went to Mexico City to begin filming *The Forgotten Village*, for which he had written the story and film script. The story focuses on a Mexican village where children have been dying due to contaminated water. A young boy, Juan Diego, seeks help from his teacher and doctors in the city, who bring medicine for the children, undermining the practices of the *curandera*, or medicine woman, who has employed traditional folk remedies to no avail. Herbert Kline, known for *Crisis* (1938) and *Lights Out in Europe* (1940), produced and directed the film, and his wife, Rosa Harvan Kline, coproduced it. Ricketts left Monterey in early June to join Steinbeck and the crew and to deliver Steinbeck's car to Mexico City. When he arrived, Ricketts initially assisted the crew in filming various scenes and helping with equipment; however, his personal disagreement with the premise of *The Forgotten Village* resulted in his short essay "Thesis and Materials for Script on Mexico." He argues, with nonteleological thinking, against the idea of bringing modern medicine to the traditional Mexican village because doing so would interfere with "what is" and would therefore undermine the culture—in this case, the people's faith in the curandera. Ricketts and Steinbeck spent the rest of their time in Mexico working independently—Steinbeck on the film, Ricketts on research for the phyletic catalog to be included in the *Sea of Cortez*. They socialized together very little.

In May 1940, Nan Ricketts moved to Alaska with daughters Nancy

Jane and Cornelia. They stayed there for a few months and in the fall of 1940 moved back to Bremerton, WA.

To Ritch and Tal Lovejoy
June 18, 1940

To R/T from Mexico City, June 18 1940:

My plans have been so uncertain and so hectic that there hasn't been much definite to write about. Now however, for the first time since we arrived in Mex City 15 days ago, we are settled in a place where I can have my desk and bookcase laid out so I can find things and feel at home.

I received a not very warm welcome—a possibility I had anticipated. Jon and Crl, Herb and Rosa, Sasha [cameraman] who is a very good man, and Mark (Herb's brother) are all living together with a couple of servants in a rather nice house. But it's a madhouse. The original plan was for me to stay there too, altho I don't know where, there are no vacant rooms. When I figured on taking Faun along, I anticipated I might be able to make some arrangement for us both to stay there, devoting a portion of my $25 per week to paying the room and board bill for Faun. Might have been able to do this if there had been good will. But there wasn't any. So I couldn't get any definite dope from that mad house. Herb seems to be very kind and capable but terribly overworked, and things can't be planned very easily down here so they never know from day to day what they're going to do. I told Herb there was no point in my staying here, and he needn't feel obligated to keep me, if I had no function. But he suggested I stay on. At first at least until July 7th elections, then later, when it appeared there was going to be declared a state of emergency (on account of war hysteria). But finally today I got my expense account cleaned up and we moved immediately into a cheaper place with some very nice Mexicans in the heart of the city. Funny, interesting place. The girls have been in the room most of the afternoon during the siesta time exchanging their knowledge of Spanish with Faun for her knowledge of English. I speak very poorly but enough to get by. Will improve rapidly since only one boy here speaks any English at all. The charge is only 7 pesos per day for room and board, we eat with the family. That's only $1.40 per day for the both of us, even at the present rate of exchange. Bought another belt, some huaraches, a fine woodcarving and a few books. There are the best bookstores I have ever seen here. I am working on and off looking up bibliographic references and abstracting Mexican scientific articles for our forthcoming book. Very fine library here at the Academy of Sciences. Was hard to find but quite satisfactory. Just around the corner. Have put in a few really good days working for the picture firm, but mostly they haven't any work for me, and—apparently (altho I may be doing them all an injustice) not much for anyone else either. They have under contract two Mexican motion picture photographers, very decent fellows. I have been going around with one of them looking up dope on witchcraft etc. Also put in two very hectic sessions photographing the fairly serious riot of university students here, which you may have read about. Got my first dose of tear gas and it's quite unpleasant but really not terrible. Sasha and Rosa and I were a team. They are good people. They can take it on the chin. I was a little scared and annoyed but went thru with it. I have been heckled so many times in my life in the past that the super heckling of being stoned and maybe a little shot at with air rifles etc (not many firearms among the students but they are certainly good at improvising sling shots etc) wasn't bad. The second time I went thru the lines bringing a tripod to where we were hidden in a balcony was unpleasant; no real danger, just the unpleasantness of being hated and reviled.

I have seen almost nothing of Jon or Crl. I am off their list; they are very careful not to invite me over there, especially when they're going to have company, and we have had meals there only a couple of times, and then when Jn [and] Crl are away or scheduled to be away. Such a feeling of coldness and hecticness that I'd prefer not to be around anyway. A new experience for me being the poor cousin. But being good natured and not holding grudges very much makes it not bad at all; also Faun [is] very easy going. We have our own fun walking literally all over Mexico City, meeting the people, riding in buses more than in taxis, and learning Spanish. I'm certainly getting to know Mexico City better than anyone could who lived in a fine house with servants and associated only with the upper crust. From the poorest peasant to the most cultured people we have met they seem to be such friendly people. Lots of sickness. Jon has been ailing most of the time. Sasha was very sick with amoebic dysentery; Herb had a bout with malaria; both (presumably) originally contracted in Spain and brought back by (I suspect) careless living in hot country. Rosa was down in bed also. We didn't take the bad trip into hot country that laid them all low but are so far the only ones not sick—says he knocking on wood. I have been having good Carta Blanca beer with each meal and often between time. Expect to leave here about July 1st, probably returning by train. See you then favor de Dios.

Faun was married to Monterey artist Andre Moreau, who lived on Huckleberry Hill near the Ariss family.

༄

To Ritch and Tal Lovejoy
June 1940

To R/T from Mexico City, June 1940

Dear people -sh'community letter. That same bottle of wine. So good. So very very good. They serve meals here as follows: breakfast 8 to 10 but nearer 10 I'm glad to say. The big meal commida corrida 2 to 3:30. Almuerza 8 to 10 at night, and again nearer ten. Very fine. All very big meals. You don't eat breakfast, you desayuna. Nearly as I can see there's no general word for "to eat," you either breakfast dinner or supper. The maid here discovered that I'm a big eater—all Mexicans eat enormously, at least they do here, and she's got to serving me two extra courses. Still that's not such an awful lot in the no. of courses that are normally served. But I had to call a halt today. Too much is too much. It was getting so I had no room for beer.

There are two or three magnificent things about Mexico City that I've never heard mentioned. The layout of the city for instance. I wish I had some photos. No one will believe this otherwise. But it's as true as I'm sitting here. At one corner there are two street signs for the same street! One says Calle Amsterdam, the other says Avenida Hippodromo. This was explained to me naively several times. It seems that before that part (the new swank part) of Mexico City was built up, the street which is now called variously Amsterdam and Hippodromo was a racetrack. That explains the fact that the street is circular. The circularness caused me a little trouble at first. The second time you pass the intersection of Amsterdam and Av. Michoacan you begin to wonder, especially when you take the trouble to go back and find that the first one really is different. Another thing that can cause you a little trouble is that not many streets have the same name for any great distance. They get tired of the one name repeated so many times, so they change it. There is a plan for the layout of the city tho, and I have a sense of it from having got into trouble so much. If you know the layout you can get places awfully fast. As in Washington DC. The rest of the guys here mostly take taxicabs and let the driver worry, but I walk or take the camion which is a species of bus, the strangest species you ever saw. A person that goes out on the street here without a city map and directory is crazy. Like going out without your hat. I saw a movie today about what hardships a group of prison escapers suffered many days on the tropical seas without any hats. I know where they'd be after the second day: dead. Oh yes the movies are fine too; there are more and larger ones than in Los Angeles. Everyone must go to a movie each day. And that's nearly true.

When you leave the table you say "buen provecho," when I sneeze in the Inst. de Biologia Library, the librarian says devoutly "salud." And oh boy how I sneeze. When you come in to breakfast you say "Buenos dias" to dinner "buenos tardes" to supper "buenos noches." When you leave you don't necessarily say "buenos noches" which doesn't have the same connotation as our goodnight, you say "adios." Also when you meet casual people or just passersby you say "Adios." Well I hope you appreciate all this. It was hard come by. I am learning Spanish fast. That is, fast for a dumb head. Here in a house where no one speaks English you have to learn it. My pantomime is getting good too. There are 39 or 72 people living in this house, there is only one toilet; you ought to see me pantomiming when I just have to go. But these people are very kind. And not too modest. They want me to go to the toilet while the maid is in there working or washing her hair or just washing some clothes in the bathtub which I suspect is used for no other purpose. But pretty clean people on the whole. There are bad fleas tho. And a few bedbugs which the little girl here calls "chinchas." I agree. But they keep after them. Our room was just washed down with kerosene. Now I understand why the maid sprayed our bed thoroly every day with kerosene in the swank hotel. There mustn't even one bedbug appear. And they are certainly well distributed here in Mexico City.

Oh yes the camiones. First here's the best story of all. (I could write a book about the facetiousnesses of Mexico City. In fact I think I am. There's the matter of the swank plumbing. What good is it when the city water pressure is shut off during the night. In one hotel we went to on the way down—bath with every room, the clerk assured us that altho for the moment the shower wasn't working, it would be turned on at 7 in the morning. But when we left at 9 there was still no pressure.).

Here's the best story of the camiones. I have been using them to get to the magnificently swank institute of biology here (you have to have a perfumed bath and manicure to get in). Costs 10 centavos. I finally found out where to take the bus to get me nearest to the library (no one knows about places outside of his own neighborhood here; even the policemen are very decent and apologetic when they tell you wrong!). (Oh yes and it costs ten cents to go to the toilet, don't let's forget that. Most of the Mexicans just don't bother. In broad daylight of a pleasant Sunday you can see some old gal squat down in the grass of the plaza at the Zocale, they really use their parks here). Well one day I went and stood there about half an hour. A lot of other people were standing there too. I can be pretty near as patient as they are. Finally I began to wonder and asked a policeman. In Mexico City a policeman is the only one to ask and usually he doesn't know, but he's very friendly about it anyway. But this one did. He assured me that it was true, I was, in actuality, standing on the very corner at which one got the bus for Lomas de Chapultepec. But "no hay estos camiones" there are none of these camiones. For the whole of the morning. And when I asked him why, he seemed to think that was strange. He considered it a minute or two with a thousand cars honking around him. Then he told me the plain truth. I think he had been instructed in the way to deal with these crazy Americans "tell the truth" they told him. He said he didn't know. And there the matter ended. For him. But not for me. After I had followed his suggestion and took another camion which made me walk about a kilometer, I asked a lady at the other end. At a little sort of a bus stop where each day I have a bottle of beer. The woman didn't understand it either. (I finally checked up on Monday, didn't go to the library Saturday). She told me there had been no buses since Thursday; she didn't know why. All this is doubly interesting in view of the fact that I did take the bus on Friday and there seemed not to be anything wrong with it. Maybe the company is bankrupt. Mexico city is a big place; million and a half population. There are lots of bus companies (the street cars handle only a fraction of the transit; everyone goes everywhere all the time here) and I suppose it's in the order of things if some of them go broke. But what do you suppose ever happened to the busses themselves; there must have been a dozen of them on that line alone?

No Americans ever take the busses in Mex City. They use taxis or their own cars. But I don't care and I have to be careful of money so I'm learning a lot that the hoity toity don't know. This bus system is screwball. But it works. The equipment is terrible, the drivers are wonderful. Every Mexican boy hopes that some day he can drive a fire engine or an ambulance. And most of them do; only their equipages are called buses or camiones. Conductors and drivers are both kids. They try to outrace all other cars including other camiones. They make simply wonderful time. Men have to get aboard or get off while the bus is still moving at least slightly. They come to a full stop however lots of times for women. The driver has an enormous mirror whereby he can cover all three or four exits in his line of vision, and just the minute you have your foot on the ground or on the bus, off it goes. You soon learn to hang on. Like flipping Chicago street cars. The signal from the conductor to go is two whistles, or if they can't whistle—some of them I imagine are too young—they say or holler "vamos" or "vaminos," "let's go." . . . The first three days Faun and I came in to town from our outlying hotel via camion, the engine broke down twice. Once the very clever driver got it fixed. The next time we had to get out in the rain. We took another bus, with a group, and the new driver was surprisingly good about not collecting another fare when someone in the bunch explained what had happened. I'm pretty well acquainted with the whole technique of bus driving except I don't think I'd be as accurate as the drivers are at almost running down pedestrians (in Mexico City, literally, the vehicle has the right of way). I've never seen anyone killed yet, but I've seen an astonishing lot of intimate relations between power poles and autos, etc.

We have a wonderful room with a balcony overlooking the busiest street you ever saw. Fine old buildings. In the university district. We are the only Americans in the whole region and everyone is wonderful to us. The saloon across the street doesn't even ask what I want anymore and I have finally worked it so the girl in the soft drink stand will give me cold Coca Cola for Faun—no water is safe apparently, so I drink beer, not a bad idea. You can get beer or soft drinks either fria or tiempo. Fria means cold. Tiempo (which literally means time) apparently means at room temperature.

But the electric lights are pretty good. They don't go off very often. So you can see how it is.

⁓

To Ritch and Tal Lovejoy
June 1940

To R/T from Mexico City, June 1940

I hope you are properly overwhelmed by my impulses in the way of writing letters. We will stay here until next Monday then come back by slow train, and literally, nearly five days. By Guadalajara, Tepic, Mazatlan, Guaymas and Tucson. Seems to be the cheapest way altho also the longest. Hate to leave. There will be real tear shedding on both sides. There are only two more things you should know first. Having recorded the Angelina Lettuce situation I find that a marimba band composed of three men, all working on the marimba, is across the street in front of the saloon where I just had a double tequila with lime and salt. Finest appetizer in the world. That is if you need one; opinion varies as to whether I do. But I have one nevertheless; and the girl brings me in the usual couple of extra courses.

And the other thing is something that happened in Puebla the other day where Sasha and Faun and Carlos and I went to film a partido revolucionario meeting. Puebla is about the 3rd or 4th city in Mexico. In the plaza an old man was selling nieves. Looks like snow flavored with lime. He had a little freezer and a bucket of water and a tray with maybe half a dozen glasses and spoons. He heaps up his glasses with the ice, sticks spoons in them, and goes around hawking them on his tray. When a Mexican wants to buy one, the old guy gives him the glass and spoon and ice, then when his tray is empty he goes back to his central position waiting for the people to bring back his glasses and spoons and to pay him. (This is a good yarn with which to explode the fiction that Mexicans are full of thievery and dishonesty; they aren't, their standard of personal honesty is far finer than that of US and I can document it). Anyway Sasha and I watched this guy for a while, admired his trusting technique, and considered what would happen if he started to sell to careless Americans on that basis. Some would forget to pay him and many would be careless of the welfare of his precious glasses and spoons. The town was full of fiesta-bent Mexicans and

Indians, terribly poor people. Most of them had never had shoes, but this was a big occasion. Rumour had it that one of the candidates was paying them one peso each to be among those present. And they were going to buy ice creams. But carefully of course, because 5¢ Mexican is a lot of money to them. But I had an idea what would happen sooner or later. And it did. One fellow finished his nieve, then took the glass back to the vendor and started to fish around for his 5¢. He looked and looked and gradually the awful truth dawned on him. He was flustered and embarrassed and tried to laugh. The old man suspected what had happened and walked away to let the guy recover his composure. When he returned the truth had to be told. The old man was jolly and kind and laughing. I couldn't hear what he said and my Spanish isn't good enough anyway. But the effect was "just don't worry a bit, old man, that can happen to anyone. Why I remember once when I . . . If you're in town again sometime and happen to see me, why fine, alright, I'll be glad for the five cents. Otherwise don't let it bother your honor's noble mind." And they parted with mutual bowings.

John left nearly a couple of weeks ago, probably you will have seen him. Yesterday we went to a neighborhood movie (they are quite different than the big cosmopolitan movies down town, just as good, more interesting, and cost 1/4 as much) with Enriquieta and Alicia here in the house. 4-1/2 hours long. One of the pictures was an item called *Via Crucis* by one J S Beck, our Jon. *Grapes of Wrath*. I hadn't seen it before. VVG. But a terribly weak closing scene I thought. If we had been willing to sit in the gallery we could have seen the thing for 20¢ Mex, 4¢ US. These double tequilas are really very good. I could be persuaded to have another if I had a little encouragement. Apparently I am the drunk of Mexico City. No one else in this house drinks at all. Via Crucis translates "the way of the cross," or the way of sorrow, sacrifice. Or so I suppose. Two people can live here in real luxury for $20 US per week. The board and room comes to 49 pesos—pesos are 5 to 1 now, were 6 to 1 when we first came down. That leaves 50 pesos per week for taxis, movies, liquor, books, stockings, trinkets, concerts. You can't really spend quite that much unless you pay extra for your meals by going to a swank restaurant. So in a pinch, two people could live here for some time on $10 per week if they walked and did their own washing. No doubt if you rented a room with a stove and cooked your own meals you could get by cheaper but we live really swank with servants to make the bed and wait on table. I think even the poorest families have two servants who must be poorer than they. No one but Tal would ask who looks after the servants. If they're young and beautiful they should get plenty of attention. So it'll probably rain again this afternoon. The first class fare from Mexico City to Nogales Ariz in pesos 109.75. Now I just thought, if we wait until next Monday it'll be election day and maybe we'll still get caught in the revolution that everyone is talking about but which I don't personally give very much credit. Has Ritchie gone to war yet. That lout that slacker that lazy is he still sitting home while his country is bleeding and dying. Wait till I sit with him and with you my fine young lady.

To Joseph Campbell
Mexico City
July 3, 1940

Dear Joe:

Your letter arrived just as I was preparing to come here, so I brought it along and have just got to it.

I have just revised—I hope for the last time—"The Philosophy of Breaking Through," and have incorporated your suggestions throughout. I found them all apropos, significant and practical.

Have also been working on the two forthcoming books, especially on the bibliography of the Gulf of Calif. acct which may develop into sort of a handbook also of the Panamic province animals. I found a fine biological library here, or rather two of them. At the institute of biology—a fine outfit affiliated to the not-so-good national university—they grin delightedly and hand me practically any book I ask for. Things I shouldn't expect to find anywhere on the Pacific coast of US unless possibly at Univ. of Calif. Life Science Library.

I may get a chance to work on the poetry essay but I doubt it. You weren't a bit too late with your pertinent suggestions. I am most grateful, for the help even more than for the attention and consideration which are most welcome. I shall look forward to your comments on the non-teleological item.

Now I think that these three essays, together with one or two more in the offing, may not be publishable until—if ever—I get well known. If Jon and I do these two things which we plan and which are well commenced, I may after that have sufficient position in order to get out a volume of essays. May be difficult tho even at that. And with a world at war, and with a US war probably coming on, may be impracticable. But anyway . . .

I think your synthesis [*The Skeleton Key*] promises better and more than anything I've heard of for many a day. I suppose it will reach only a highly restricted audience, but it seems awfully important to me. It ought to be done. Must be awfully thankless work, considering prerequisites, difficulties, and limited audience, but I hope that something will stimulate you so as to keep you at it. All art is in one sense dedicated to the recognition of the all in everything, altho it's not generally recognized. The diamond is universally sought in all fields as the symbol of the most persistent and permanent thing in the world. It has no other value except in the crafts. Oh by the way I use the term holistic in the sense that Jan Smuts considers it in his essay "Holism" in XIV *Encycl[opedia] Britannica;* may stand a definition tho: a light; the integration of the parts being other (and more than) the sum of the parts; an emergent.

Lots to do; home next week, and much work before then. Good luck in your work, and best regards to both of you.

> Ricketts's final remarks about holism are in response to Campbell's comments on "The Spiritual Morphology of Poetry." In his letter to Ricketts dated May 19, 1940, Campbell wrote: "What is the meaning

of 'holistic tenderness?' Holism is a philosophy which declares that the determining factors in evolution are entire organisms and not the constituent parts of organisms. You may have a big idea here, but the statement requires clarification."

⌇

To Herb Kline
1940

Dear Herb:
I had a great deal of trouble about Jon's car, and it's only by the grace of god (in the fall of the dice of chance), and my own good nature and relaxation and maybe a knowledge of the way the Mexican mind works, that I got thru at all. When I showed up at the border with a tourista stamped "automobile" they just went up in the air. Asked me where my car was. They suspected it had been sold in Mexico without payment of duty. I told them only the truth and no more than I had to: that it wasn't my car, that it belonged to a friend of mine who was there at the time I got to Mexico City who had gone down by plane and returned by plane but would fly down again, and that I had made proper affidavits at the time. First they said I'd have to go back to Mexico City and get the car. Looked for a moment as tho they meant it. They even held up our Pullman car and wouldn't open the barbed wire gates to let it across until that had been settled. Well I felt good and I liked the guy I was dealing with, and he was hot and lame and a good fellow trying to do his duty in a difficult situation. And he so wanted some good strong mind to lean on! He didn't know what to do. He tried to find his superiors but everything was closed up in the siesta hour when finally he discovered that he'd let an auto-less "auto-tourista" almost slip thru. So I said it was perfectly alright that if they wanted me to go back to Mexico City I was in their hands and I felt alright about it. In such a case I would have called you up in attempt to arrange things so I wouldn't have to go. But they finally took my word for the fact that everything was on the up and up. And so I got thru only into a mess of lost baggage. They succeeded in losing my

precious bookcase. Had many of my notes in it. But fortunately the most important I had put in my brief case.

So anyway some of you guys who bring Jon's car finally thru, either will have one hell of a time or else maybe they'll welcome you as a bringer of good tidings. I hope that if the guy who let me thru has lost his job as a result, your showing up with the car will reinstate him.

Card from Jon and am writing him tonite; haven't seen him or Crl yet.

Oh yes and when it comes time for you to go thru the line bring along plenty of evidence of US citizenship; they certainly have the border tight shut. Hello to all.

To Nancy Jane and Cornelia Ricketts
July 1940

Dear Nancy Jane and Cornelia:
You sent that book on Lincoln with rare insight. Not only is it one of the two great Lincoln Biographies (the 6 volume Sandburg is the other) but it's a great book. And about a great man. And about one of my two personal heroes (St Francis is the other; I just bought "The Little Flowers" etc in the Everyman Edition). "For he was a citizen of that far country where there is neither aristocrat nor democrat"—the country in which everyone, consciously or unconsciously, is more interested than in anything else in the world.

I have just only rtnd from Mexico City. Jon and Crl came back by plane, I by train. I had almost 5 weeks work in the library of the Inst of Biol there, and got a good grasp on the bibliography for the new book. If only we had some illustrations, it would be a dandy. But it'll be fine anyway.

I hope you're having a good time, and know you are.

By mistake 2 checks were sent you last month. That now, acct to my records puts me two months up ahead. Be careful of the money, lord only knows where more of it is coming from. If we can only get these books moving, royalties will look after things for [a] while—I hope—but that's in the future and the not-too certain future. Business is dull; for one reason because I have been off writing books instead of tending to it. But of course in the long run, if I can in the meantime hold on, the results will justify that diversion. Will you be coming back for continuing school at Bremerton, or will you stay up there? Best love to all

Ed

At this time, Nancy Jane was sixteen and Cornelia was fourteen years old.

To Nan Ricketts
July 30, 1940

Dear Nan:
I got your airmailed letter Saturday. Mail is otherwise awfully slow from Sitka.

There is very little money available, but I think a postdated check for Sept. 1st will be perfectly safe. Believe I am still a month or two ahead. Will note it on the check. In addition to the remittance I sent from Mexico City when I got your airmail letter down there, Cornwall sent one from here. Mine was sent to Nancy Jane, his to you.

Things are going along so-so. I have been working hard on the new book. My part of it is really pretty well up-to-date. Jon plans to tear into his part [in] October and work right thru until finished possibly in the spring. Unless the war intervenes which it very possibly will. There's a tremendous lot of work to it, especially from my standpoint. I haven't Jon's positive genius to go on, so have to make it up by plodding hard work. And then the very nature of my contribution involves long and studious work. I wrote up the original notes based on our travel experience, from which Jon with his fine thinking and writing will work up an interesting and I imagine worth-while account. Then I am sorting and sending out to specialists the animals we collected. When those reports are at hand they must be collated and analyzed and worked up for listing in the scientific appendix. Then I will work up a bibliography (almost done now) and a list of collecting stations and what we found there. But it will be a worthwhile piece of work if we ever get it done and out.

Edward is coming along alright. He was never dangerously sick, but was in some distress and for a while it was nip and tuck for an operation. I hate to think of this new crop of bills. And just after I got every cent of Dr. Hoyt's old bills paid up! Ed by the way has grown so that his feet were too big for the moccasins that your folks sent. However that wasn't an unmixed blessing. They fit me perfectly!

Yes the SE Alaska region is pretty rainy, altho this month should be pretty clear. I hope you get lots of Gonionemus. I didn't expect that any would show up this year, so haven't advertised them, and because of non-deliveries in the past few years and because I didn't advertise them very heavily in the recent past, we haven't any orders whatsoever. But I will get out a few letters and see what can be done. Get what you can anyway. I can sell them eventually. Main thing is to have them on hand before the orders arrive. Keep on the lookout also for brachiopods, the clam-like things that are attached to low-tide rocks by a little stem. They are needed. Some cucumbers and other lot of Pleurobrachia will have to be got in the next year or so too, but I'll probably have to tend to that myself, acct. difficulty of anesthetization and preservation, danger of using osmic acid etc.

Peg is permanently separated from Jim [Fitzgerald]. He finally divorced her. I had quite a nice, slightly sad letter from her within the last year; she was asking about Jim, if he was happy etc. The Cohees were in here once during the past year. They are in Los Angeles. Just stayed the evening; we had dinner here with some abalones John caught illegally. Seems that Alice (this according to Jon) had another bout with her TB and went to Arizona for awhile. I don't know how they're working out. Believe he is on a newspaper there.

Jon turned over to Ritch a $1000 cash prize that he didn't need. Ritchie then quit Holmans where he had been working for more than 2 years as [advertising] mgr., and started to finish his novel, which Pat Covici (now with Viking Press) promised to publish. Now they are at Jons and Crls for a few days. I came back from Mexico City on the train, leaving Jon's car there. He flew down. I was supposed to do a little work there on the movie they're making, but there was very little for me to do except once or twice drive a car around thru mobs of rioting people, where a Czecho-slovakian motion picture photographer, Sasha, got some action pictures. Got a pretty bad dose of tear gas once, and stoned a few times, but not shot at. On election day Herb left me with Jon's car while he and a Mexican cinematographer went up in a building to get some shots of one of the bad polling places. Told me to leave the car and seek refuge in a doorway if things got hot. So I did. Then pretty soon he came down and said "Ed you dam fool get away from there" (in the meantime a great group of Mexicans with green banners etc had come into the same doorway. So he pointed up to the roof opposite, and I saw there was a machine gun mounted on a tripod and trained on the very doorway in which I had taken refuge. But most of the time I worked on our book.

Well, I have too much to do. I spent so much time taking that Gulf [of] Calif. trip and working up the animals and working on the book (which seemed a wise thing since if it ever does come out it will bring me a good income for a year or two) that things at the lab sort of went to pot. Few orders and no money. So now I have to do the best I can to put them together again, and sidetrack the book so as to get money enough in to earn a living. Cornwall who was here an hour a day was a good honest man and he tried but he didn't savvy and I really couldn't expect him to learn such complicated work so rapidly.

Take good care of that cold, write when you get a chance.

> The illness Edward F. Ricketts Jr. suffered from was kidney stones, but he did not have an operation, as doctors were able to dissolve them (Ricketts Interview, June 28, 2000).
> In 1940, Steinbeck was awarded the Pulitzer Prize for *The Grapes of Wrath*. He gave his $1,000-dollar award to Ritch Lovejoy to work on a novel that was never published.

To Herb Kline
August 17, 1940

Dear Herb:
You did mail the letter. More power to you. And I received it thanks to the pleasant sort of cooperation btwn Mex. and Uncle Sam that the best in the postal service brings out. Jn is here now, I told him about the need of writing that letter re[garding] the car. And he promises to do his best. So a pleasant time!

I finally got to replying to Mark [Kline's brother]. (Note that this letter goes forward the same day I receive yours!). Yes I left all the papers with Mark incl[uding] the pink slip. In, I think, two installments, with Crl superintending the 2nd. I handed in the first papers which presumably included the permit right away, the time the insurance man was expected there, and right before you took out Mex. insurance on Jon's car.

Ritch and Tal are having a baby. Or are in the process of having one which you'll grant is still more pleasant. And Crl promises to give the child a scholarship in the school of hard knocks. And I'm doing a little like-mad writing myself. If we ever get out that gulf book it will be a very nice job. Amoebas is sometimes hard jobs to cure [*sic*]. But keeping them on the run permits a person to go on about his life. My sister has had the bugs for many years; takes a course of yatrin about once a year or so and is alright the rest of the time. The curandera thing will have been fun; I'd like plenty of time and opportunity to chase down a thing like that, collate and document the results, keeping on until there was a sense of holistic understanding; a good deal of real information could come out a thing like that; there's almost always some reality, symbolic of not factual, underlying that sort of stuff. We heard something about the C's [reference unclear] policy up here. In Fact. But it's significant also to read indices of facts as they are, despite their etiologies—which are always disputable; people or votes or winds don't always go the way they should go. They go the way they go. Which may very easily be having you and all of us—but I'm over 43 now and just let this good democracy fritter around for a while more and I'll be past the age—carrying guns around. I think with Rosa that the medical corps is the answer for me. Well, better to stop and drink rum, and hope you've dug up those papers. You'll probably hear soon from Jon about the car. Hello to all—is Rosa all OK and over her sore throat?

> The "curandera thing" Ricketts mentions in the third paragraph was possibly his intention to study the practices and/or methods of traditional curanderas in an attempt to achieve a better "sense of holistic understanding." Despite his personal disagreement with Steinbeck's *The Forgotten Village* project, Ricketts was clearly intrigued with the dynamic that existed between the villagers and their healer—the curandera—exemplified by their unwavering faith in her abilities. Whether he actually pursued such a study is unknown.

To Toni Jackson
October 1940

Dear Toni: Lots of happy tender thoughts about you yesterday, but no time for writing them. We stopped at the ranch, had a bit to eat at the lab, then stayed for a while at R/T looking after the baby while they went out.

Crl doing alright. Was concerned to make out what a beautiful relation they had had, which I can't see—and said so; then with how much she loved and has loved Jon—which is more than she has loved anyone else certainly, and therefore the only thing she has to judge by. Therefore it is true within herself that she has loved him and does. But as I use the term, I still don't see it. I think that all the love she has is Jon's, and that she can love; but in a deeper sense I don't think she has ever really loved anyone. But I am subject to grave miscalculations when dealing with the type that I call "bargaining" as contrasted with the type I call "participating," and I don't know in such cases ever whether I'm being inwardly just or not. My own inward picture of the situation I have finally worked out, and it's dynamite. Probably I shouldn't say it even to you, but making outward an inner thing is a good idea and here it is. That in the relation between Jon and Crl, Jon seems often to be (objectively) "wrong"—that is in disagreements between them Jon often has his neck out and Carol is sound. But that the more important fact is that Jon is a far more consequential, deeper person. Or rather that he lives on a level higher (to me) than hers, or that his life is larger, his horizon greater; however you want to say it. Which isn't all-important, because Gwen, who seems better adapted at the moment or at least with whom he gets along well at the moment, is also a less person than Jon. Probably less than Crl. But in addition to Crl being less than Jon, there is bad blood between them, and I think it's truly on both sides despite Crl's claim that she became what she has become—a hard, brittle sophisticated wise-cracking inward hiding woman—in response to what Jon said he wanted; and the circle is going around the wrong way. With Gwen, who is maybe even less a person (but more malleable in a womanly way, more womanly, more earthy) the circle is going round right; all the things they have together so far are positive, whereas many of the things Crl and Jon have together are negative. So no wonder Jon gets deep peace with Gwen whatever she is—and so far she seems to be loyal and suitable—and doesn't have it with Crl even tho she may be a greater person. And when on top of that you see that Crl is hard to get along with and herself needing training and a strong loving hand, it's easy to see what a strain it would be on Jon who himself needs loving tolerance and firmness, not criticisms and abjectness.

We had quite a time coming down. Crl wants to make out that Jon has often been unjust, which I guess is true; that he has been very strait-laced about sexual matters with other women—which good god almighty certainly is not so and I have indicated or hinted it to her but how can you come out and tell her a thing like that even tho she's building an important theory on her inadequate knowledge; that they have had a beautiful relation which I don't think is so; there's been some good times (looking at it from the standpoint of an outsider) but an awful lot of strain and bad times as compared with R/T etc; that Jon is more parsimonious about money matters than she is. Which last may be or may not be, I don't know. Both of them have a curious relation with money for rich people (but I guess most rich people use their money to get more or merely to hang onto what they have, even tho in their lives they can't possibly use it up). Jon or Carol or both have been in big ways generous with me, altho in small ways I often find myself paying for things that I can't afford, to save them the trouble of cashing a check etc for something they *can* afford. But when I remember how upset Carol got when, right after she had showed me a check for over sixty thousand and said "the money's coming in so fast I don't know what we'll do with it" she read further on in the statement, and discovered Jon had drawn $500 when he was there. And how sad she was when I told her I thot Jon's car was getting about 13 m.p.g. And again last night when she said she thot she'd get some Chintz curtains for the new house; that they were quite cheap. And how much money it was going to cost to fix up the present house, that Jon wanted fine bathroom etc, which would alone cost as much as he told her to spend on the whole rebuilding. So I guess the situation is that the mothers of both of them were seared by a dollar bill before their respective children were born. Of course it's hard for people who are naturally parsimonious, or who have had carefulness thrust on them as an adjustment to poverty, to suddenly change around. The old habits, especially if they are instinctive, are strong.

So I've been on a spot, trying to insist what is true—or rather trying to differentiate what I, due to my characteristics and bugs, think is true, from what is more objectively true, and insisting to Crl what is objectively true and straining out just what personal—I think is true—and at the same time letting her feel my love and comfort. Most people, when they love and comfort someone else, do so at the sacrifice of true thing, and so help a person to build up a still further structure of untruth. And I want to avoid that. Well, I'll probably be in the doghouse anyway. Carol will be sore as hell if she is influenced some way she doesn't want to go by anything I say, whether it's true or not.

Ed Jr just came in with the news that Jon took Tiny down to LA with him. Because Tiny's brother is a fellow student of Ed's and has been driving Tiny's car until he comes back, today I think. No word direct tho. Everyone is stewed up and I seem to be in the center of things, with everyone from Crl father to Esther talking to me about it. Wotta poise I have to try to maintain who am not myself very balanced.

Well honey all this letter to you just about that. But alright. Because it gets my own ideas out and organized, maybe so I can see the bugs in them too. See you soon. Am enclosing a couple of things you wanted, plus that $5. If I don't write a check out to you specifically lord only knows when you'll get it. See you soon.

> Toni Seixas Solomons Jackson (1911–) married Ben Jackson in 1933 and gave birth to their daughter, Kay, in 1935. After their divorce, Toni met Ricketts in 1940 through Virginia Scardigli and later moved into the lab on June 10, 1941. She and Kay lived with Ricketts until Kay died in October 1947 of a brain tumor.
> Steinbeck met Gwendolyn Conger (who later changed the spelling of her name to Gwyndolyn), a Hollywood singer, through friend Max Wagner in 1939. Soon after Steinbeck completed *The Grapes of Wrath*, his marriage to Carol deteriorated and he fell in love with Gwyn. In 1941, Steinbeck and Carol separated, and on March 18, 1943, their divorced was finalized. He married Gwyn eleven days later.

To Gwyn Conger (Steinbeck)
November 27, 1940

Dear Gwen:
Long time, sure enough. Glad to hear. I wrote Jon at once about Max [Wagner]. I fear there isn't much to expect in the way of his stopping drinking, even under pressure of friendship or threat of illness but I agree thoroly with you that those things are worth trying.

I may be up to Los Gatos Sat. PM. A girl friend has been coming down from the city to spend her week ends here; I imagine that we will collect in Santa Cruz on the Saturday 5:30 PM tide and we may run up there afterward for awhile. In any case I'll see Jon sometime next week; can mention the Maxie thing in more detail then. No excitement! You can't stop that for Max. He eats it up. That's his food and life. I imagine.

My canning is never finished. I have just been working over some animals that not even your mother would believe. She just couldn't. They aren't there. And furthermore I've cut down a little on drinking myself. Why even this minute I have only one glass of wine before me, and that's been filled only twice. And before that I had (in this order) only a can of beer, two drinks of rum, and my share (split 3 ways) of a quart of beer. So you can see everything is in order. Your mother would see that if she were here. You'd see it yourself Gwen.

It's on account of my having a cold that I've stopped drinking. And because it's too expensive. And it isn't good for a person's health. Everyone knows that. You wouldn't have me drink myself into an early grave, age 44, would you? Yes you wouldn't, like I would.

I have been working on the book at a great rate. It's lining up beautifully. Jon has started to put good licks into it now too. Incidentally, if you want to know what Ritchie's friend is up to. He just pretty nearly kicked a book out of Tal's lap.

Yes Gwen, the beard's still on deck, but my valences, work, play, feminine and otherwise are pretty well taken up now.

Sure, I will not forget you Gwen. If people have had between them even for once that good honest spark, there's always something there. It can be forgot, alright, it can be got out of touch with; but the thing itself is always there. You will find me a good friend to you, I think. Tell you what, come up here before you go away; I'll drive you up to Jon's. We'll see if we can get Crl in a sedate mood and then just devil the life out of her. May be we can devil that wicked old Jon a little bit too. Worth coming up for? Good love to you, Gwen, have fun.

To Toni Jackson
December 23, 1940

You nice Toni: The best thing I do Monday evenings, just back from SF, is to write you. I am so glad to be doing that now:

I got Ed's record finally at the White House; cooperation was very poor there. The clerk was surly, uninterested and didn't know the material. For a while I considered whether I would report him to the head of the department, or, lacking cooperation there, to some one higher in the store. But then thot of Christmas shopping, and the type of dumbbells who take up a clerk's time inconsiderately, etc, and let it go. But I did want that record and refused to be put off just because I didn't know the name of the performers, so we finally worked it out. But it was pure persistency on my part.

Jon [Cage] and I had a very pleasant trip back. We worked out an understanding, almost a statement of our differences of viewpoint. It involves, as I suspected, a real honest to god fundamental, a right or left turn up the steep mountain, and surely involves in culture that same primitive cleavage apparent now in government as an individual or a communal point of departure. And he represents the probably oncoming thing. Inspired geometry is a good term.

A square or a black line is more nearly the same for all people—therefore a great leveler—than a folksong or a picture of a cow or a Shakespeare sonnet, or more even than the tones or words etc of which they're composed. But all my tendencies are towards "meaning," while his are towards "organization" as such. I regard content as primary, he form. It's more than the old controversy, it represents actually a fundamental divergence (altho I still think it's the mountain that's of deepest importance). And his is unquestionably the purest thing. He regards all sound (but especially sound devoid of previous (traditional) meaning) as the subject matter for organization in that discipline which is music; whereas the former tendency has been surely (but in music less probably than the other arts) to regard meaning, what the artist had to say, as the prime mover. The way he said it also was important, and a matter of discipline, but the important thing was that he had "something to say." The newer idea, of "inspired geometry" concerns itself with pure form, with pure building blocks themselves originally devoid of meaning. The meaning essentially attaches to the new form as an incidental but probably essentially sequential, by-product. But the thing is pure in itself "let the chips fall where they may." "It's what it is." Pure. And that's the point of contact I was interested in—since the pureness of a thing is what I like also, and therefore a common ground. His type of "word of god made man" works out thru form, a meaning attaches thru association for whatever it is, wherever it goes. My type (and the more conventional types from which it stems), derives thru what it has to say, and then—technic secondary and a result (altho an essential one and one that improves with work)—laboriously takes the form most suited to it, or the form in which I am most competent. It finally boils down to a matter of emphasis. And it is, after all, a vitally fundamental difference, but it's a difference of way, or of "Tao" whereas the thing of central significance—you just can't say it, it's so pure and refined—involves the central tao. It's the mountain that's important, rather than the way up it, which is also however important as being part of the mountain.

Well, all that occupied most of the trip home. That and eating. It started out from the fact that a fellow in California Arts and Architecture thot that John was doing interesting and (I suppose more importantly from a publicity standpoint) spectacular work, and that a writeup would be good. So he wrote it up, based on a credo or manifesto that John gave him. And the guy missed many of the significances entirely, and got others of them factually wrong, which is annoying to a perfectionist. So John was trying to write him a letter suggesting a revision of the article (altho I think that discipline and organization, based on an understanding of the facts, are the prime necessity in the editor's mind). And he was having trouble wording it. And thot we might get some illumination on the whole thing on the way down. Which we did. Or which I did anyway. And now he hopes he can work with Glotzie who is fairly articulate but more important, who understands and herself stems from John's highly rarefied viewpoint. All the actual discussion came from our consideration of my definition of poetry. Which as I knew, didn't include the Stein-Pound-Joyce, etc, school, nor the surrealists, etc. And of course any definition that would derive thru unaided-me, would consider content as primary, and so wouldn't be likely to include them. A new definition is needed, if there is any honesty and discipline in that group. And there surely is in at least some. And some damn good workmanship. John has the same respect for fine craftsmanship—honest and disciplined craftsmanship—that I have. He himself accomplishes it. And others in that group surely must be that way. And since people like John and Glotzie, who are in it and of it, are yet very conscious and articulate, an explicit statement ought to come thru them. Lord knows there have been few enough explicitations from that bunch.

But I got some respect for myself, even, out of talking today to Jon as a proponent of a thing I am not part of, but that I believe can come and may come. Because I think more than ever that the important thing is the whole picture. Not his bias nor my bias, his way or his viewpoint or mine, but all one can see thru or with his essential framework as a vehicle.

Interestingly, I got new incidental light on the collectivist movements, on the essentialnesses of Hitler and Stalin as expressions. A no doubt coming thing, altho possibly—probably—destructive of our way of life and of us personally—people like John Cage and I, altho he doesn't believe it.

Well, I had something else I wanted to tell you. Forget now even what it was. But alright; letter already too long. Alright again tho, since my writing to you is at least an expression of my thinking of you. . . .

>Ricketts was greatly interested in John Cage's experimental percussion music. As evident in the letter above, the two men enjoyed exploring how their opinions differed and often discussed these differences philosophically.

Ed Ricketts in Chicago, prior to his 1922 marriage to Nan.

Albert E. Galigher, Ed Ricketts, and their wives and sons in 1924, when the two men opened Pacific Biological Laboratories in Pacific Grove, California.

The Ricketts Family, mid 1930s. Left to right: Ed Jr., Nan, Ed Sr., Cornelia, Nancy Jane.

Ed Ricketts working with specimens collected in Sitka, Alaska, during a 1932 trip with Joseph Campbell.

Joseph Campbell (center), Ricketts (far right), and others on Calvin's boat, the *Grampus*, during a 1932 collecting trip to Sitka, Alaska.

Ed Ricketts in Sitka, Alaska, 1932.

Ed Ricketts with Xenia Cage in Sitka, Alaska, 1932.

Ritch and Tal Lovejoy playing in the backyard of the Steinbecks' house in Pacific Grove, California, mid 1930s. Their game in this instance was dubbed "Eviction Brings Sorrow."

Ricketts and Tal Lovejoy in the backyard of the Steinbecks' house in Pacific Grove, California, mid 1930s.

Pacific Biological Laboratories, 1930s.

Pacific Biological Laboratories burning during waterfront fire, November 25, 1936. Photo originally appeared in *Monterey Peninsula Herald*.

The *Western Flyer* returning to Monterey Bay after the Sea of Cortez expedition, April 1940.

Left to right: two Mexican hunting guides, John Steinbeck, and Ed Ricketts in Puerto Escondido during the Sea of Cortez expedition, 1940. Photo courtesy of Toni Volcani.

Ed Ricketts, circa 1941.

Ed Ricketts's mother, Alice Ricketts, Ed Ricketts Sr., and Ed Ricketts Jr., circa 1941.

Toni Jackson and Ed Ricketts, mid 1940s.

Ed Ricketts and his brother-in-law Fred Strong, circa 1943, in the living room of Strong's Carmel house.

Ed Ricketts and Toni Jackson, Queen Charlotte Islands, 1946.

Toni Jackson, Kay Jackson, and Ed Ricketts Jr. at Pacific Biological Laboratories, circa 1946.

Ed Ricketts in Monterey a few months before his death.

1941—1942

To Toni Jackson
January 22, 1941

Wed. evening

Dear Toni:
Several things.

First: here is your five bucks. And thank you.

Second: Gustav [Lannestock] just called. We are to go over there Saturday 7 PM, if we will (and I accepted for you) for smorgasbord. Will drive over in a few minutes to tell R & T.

A while back you offered to do any typing that might help out on the new book. If you have time now and want to do it—but look Toni now, don't be foolish and undertake anything that's going to be a burden—I should like to get a clean copy of the latest draft of Non-Tel[eological]-thinking for Jon. He wants to use some of the ideas in the new book; his copy is the old one you read at Virginia's; there have been many revisions since; and it will be handy if he has it for work soon to begin here. Then I can keep my copy inviolable; otherwise it always gets away and I have nothing to refer to.

I was glad to write you Monday; thot of you a lot; as you are often in my mind. Remembering what you said about being sentimental. Honey don't you worry about such things. Be yourself. Sentimentality has an underlying nobility, and the only way that potential can be developed is to work thru to it. It may be necessary a means to a good end. It's only the sentimentality itself that's sophomoric; the thing that is farther to it and which may emerge from it is good, but that thing will atrophy or at best be left still merely as an unrealized potential otherwise.

Finished your "Heart is a lonely hunter" and passed it on to Tal. Now your "Stone of Chastity" is ready for return, having gone the rounds. I think that the universality of projection and the tragic necessity of it is the chief point that the Cullen gal makes; and the projectee rarely even knowing what it's all about. But if that's the way personal development proceeds, then that's the way it proceeds. The Orientals apparently regarded that as a trap, and think they discovered means for escaping from the trap.

I finally got the Hemingway thing worked out in my mind. This is the way it seems to me to be. The clue is in the title as it is in many significant writings. He didn't say "To have and to have not." He said "To have and have not." The point of the whole book is surely, as you said, (but as I might have tended to overlook), in the words you mentioned. But the idea is that this man (theoretically representing the "have nots") who so richly had, at the same time "had not," in the sense that the "haves" (in traditional sense), actually "have not." The real tragedy of the thing is that Harry, who so richly "has," also wants to "have" in the more traditional sense; and that search proves his undoing. And in a real sense the book is a contrast between the "haves" (i.e. Harry), and the "have nots" (the wealthy guys and girls on the yachts); but not at all the way the communists would figure it; not at all in a class consciousness sense. "I couldn't go to the funeral. But people don't understand that. They don't know how you feel. Because good men are scarce." They had. "You just get dead like most people are most of the time." Most people haven't.

It's a story of the tragedy of blindness. That rich people who have don't use their wealth except to make them miserable; or anyway don't see the blinding light of the richness (that people like Harry and his wife have), and the tragedy there is that they don't even search; they don't know they lack. And that Harry himself isn't satisfied with his "have" which is a have on a plane far above that of the rich men; and so he goes searching too. And he thinks he "has not" altho he "has" far more richly than the rich men who think they have. Finally, the last paragraph in the book "A large white yacht was coming into the harbor and seven miles out on the horizon you could see a tanker, small and neat in profile against the blue sea, hugging the reef as she made to the westward to keep from wasting fuel against the stream." That contrast, but only on the most superficial level (as superficial things may be symbols of deep things), persists clear thru to the last.

Have been so steamed up the past few days I find it hard to get out of the field of thought and inwardness (what I heretofore regarded as creativeness) into the ordinary pattern of external work. Everything has a potential for falling in place nicely; I can take any given problem and "work it out." The potential for falling is "n," and I can take any one of those leads and work it out. Of course that doesn't mean that you can work out all the leads, but you can work out any one of them. Funny.

A new lot of identifications back, this time on worms. There were some 41 species in that lot, 2 of them new to science. And now I have sufficient data to start reading on that group. Another big job, but not nearly so entangling as one or two that I've done, since Dr. Hartman sent me her papers to date, which unsnarl most of the things I ordinarily have to do. I usually, in my haste, waste more time trying to do things by short cuts than if I had used the approved old method at first, but this time will just laboriously make up a card catalog bibliography from my bibliogr. notebook. And a big lot of stuff finally ready for freight shipment to the Natl. Museum. Which will leave me only half a dozen groups still to sort, pack and send. Maybe we can take the balance of the mollusks to Dr. Berry when we go south. He is at Redlands; a little out of our coast way; but it will save some work not having to pack them, and I can jack him up personally which is more effective than by mail. I know what he'll do: promise "tomorrow," just as I would under the circumstances. And then not find time to do it.

Nice sense of accomplishment that comes from figuring up good things: with the $11 payment today I have only $90 more to go on Nan's old Dr. bill. All that's left of nearly $1500; doctors, hospital, nurse, drugs, incidentals; wotta load. We used to have a superstition that we mustn't pay that last five or ten dollars on the family doctor bill. That then another bill would start to pile up. And it worked out last spring that way alright. Within a few months of the time I had paid our present doctor his last dollar, Ed was taken sick, and the bill is back up to over a hundred. But I do like to get them cleaned up; maybe sometime they'll stay that way for a while.

Well honey, still more animals to sort. Come down soon to the welcome that always awaits you here, to your good warm welcome home.

>   Toni was, in fact, employed for a time as Steinbeck's typist, working on *Sea of Cortez* and other projects.
>   Dr. Olga Hartman published extensively on worms, specifically on the subclass Polychaeta. Fourteen of her articles are included in the appendix of *Between Pacific Tides*.

To Nancy Jane Ricketts
January 24, 1941

Dear Nancy Jane:
You are certainly going right along in music; thank you so much for the clipping. Well well. And getting her name in the paper for singing. Next you'll be doing Monteverdi and Palestrina as written, modally. For that music, object is to keep away from the piano and the piano key.

I haven't had any new music for quite a while. But for some time I have intended sending you some of my records anyway, and now I have got them ready to send. There is one real gem and the record is in perfect condition: the Russian credo of Gretchaninoff. When I can afford it I will get myself another copy, Victor records are about half price now. The other things are old and not worth much, except for the Gregorian which is the very best of the Gregorian singing. The record however is a little scratched; it was got for half price at a 2nd hand record store sometime back in SF by Dick Albee, who gave it to me. I hope eventually to get the whole album, so take delight in sending that to you against that time. The other two are old Stravinsky's; you will remember them probably with some pleasure, having in the meantime forgot.

Also I am sending a Thoreau; the one I had and have been reading at. R & T since gave me another copy, and in it I have transferred my markings, which serve at various times as notes for various oncoming essays. Sometime, if the book that Jon and I are writing is ever completed, and ever published, and if the other one we started ever gets completed, I may be well enough known so that I can publish a volume of essays; three are done now.

Don't know if you'll like the Thoreau, but I know you will the music.

It seems more important, now that I am pretty well strapped financially, to pay the electricity and phone and water, and so keep going, rather than send the money up there and get cut off and have to stop business entirely. I sent your mother two sets of checks in November, one for Nov. 1st and one for Dec. 1st. I owe for the Jan. 1st, and I don't have the money. After I go to the PO today I may find something there, in which case I can enclose a check with this. It looks as tho I'm going to be financially strapped now for some time; maybe not intensely until midspring, but things look pretty slack. However, if I can keep going until Jon and I get this book out, I will have ample funds from royalties; so I can keep my fingers crossed with that end in view. Wondering in the meantime by what miracle you folks can get by. Still I'll probably send a good part of the money due, and you folks are probably earning some up there; it's a good boom town I hear. (As Monterey is also). Ed and I have it fairly easy; we live here at the lab, and as long as I can make interest payments on the mortgage, and pay the public utilities and a few groceries we can keep going. I figure that another year will do it. But that year may be pretty hard on all of us. Main thing is to keep up morale; I'm even taking s.l.o. [shark liver oil] now chiefly for that purpose. And to eliminate chance of further colds. That one that I had slowed up my working for a week or so.

Ed got only a B in biology. I thought that was amusing. In some practical and field collecting ways, he undoubtedly knew more than his instructor. Now soon you and he both will be graduating. Well, love to you. I will write to Cornelia soon, to whom my love is sent, if ever I can think of anything to give her that she'd like. What do you think? Preferably something I have here, but I could probably dig up a dollar or two to spend in cash if there's something specially she'd like. Sort of as a belated Christmas present, or just as a love gift.

To Peg Fitzgerald
Early 1941

Dear Pegs:
Glad to hear. This typewriter that keeps skipping spaces! Jon and I sidetracked our first book in favor of a second based on the Gulf of California trip; and that also is just in the process. I have my share of it maybe half done; Jon expects to start licking his share in another month or so now but I imagine there will be delays on the parts of both of us.

We took a chartered purse seiner down into the wilderness of the Gulf of California; Jon, Crl and I, and [a] crew of 4. We also helped work the boat which would otherwise be undermanned. Fine trip. No hardships despite the excessive heat and the wilderness. Put in at only 4 ports: Cabo San Lucas, La Paz, Loreto and Guaymas. I worked hard enough so that I lost 5 pounds despite eating like a horse and drinking all the good Mexican beer that I could lay my hands on.

And after that I went to Mexico City for a couple of months, working most of the time in the very good library of the very good institute of biology which is a more or less autonomous branch of the not-so-good national university. Laid the foundation there for the bibliographic research on the new book which I have been continuing since at UC, Stanford, Hopkins and in my own library which is slowly building up again.

All the "pure" (i.e. non-expedient) things are likely to be in a period of doldrums during the anticipation of war, actually more than in war itself; altho there it'll be hard enough Lord knows. And subsequently worse, if as I suspect, authoritarianism will be creeping in on us no matter who's president or who wins the war or under what name it emerges. But I can remember a little gleam of hope from the last world war. At first people had that sense of "eat and drink for tomorrow we die"—which has some good values in it also insofar as it involves a relaxation and emergence of one's own personality as opposed to the usual duty persona. But then people got to realizing that there was still work to do "not unbecoming . . ." and that fairly deep idea was expressed in the superficial phrase "business as usual." Because it's surely true that along with wars and alarms of wars people have their own individual integrities. And if a man at peace is a scientist or dancer or writer or businessman, he is that also during war if he keeps his head up. At first, in the current too strong to swim against, everyone is spun over and over and loses his head; some stay under, but many emerge and swim with the current keeping their head up and maintaining some sort of orientation toward an original or a changed objective. Well, when I realized all that this time I felt better. I notice Jon has seemed also to go thru something of the sort. At first he was so excited about the war situation that all his plans including ours were upset. He'd say "I'll do thus and so unless the war intervenes." Now I notice he plans to do thus and so insofar as he is physically able, despite war or sounds of war.

My father and I were registering in the last draft. If it goes again to 18 to 45, Ed Jr and I both will be registering in this one. I am secure from drafting unless I choose to go in; and my false teeth and bum feet might even so keep me out at age 44. Last time I went willingly at a naïve level of patriotism which had in it merely the shadow or symbol of a good thing. When I found that the army then (after I had been transferred, at my own request, from detached service in the medical corp, back to the infantry) was full of graft and petty inefficiency, at first I did more than my share and was bothered; then "soldiered along" and did less than my share, tying in with the petty graft. Then afterward for a long time, I should have resented going again; I could have classed myself as a conscientious objector. Now I am back again where I was at the start, but "easy" and relaxed. If necessary, I would go in again gladly, but realistically this time, knowing more deeply how things and people are, and shorn of a whole group of unworkable ideals; but on the whole happier than then. Which is the saga of one eddie, doughboy.

Ed Jr only is with me. We are making out very well. I discover after many years that I am after all a good parent; a very good one. Which is a nice thing to realize. And more non-possessive about my children at least, than most people are. And I find strangely enough that after all his alleged (by me) inconsideratenesses, Ed Jr turns out to be the most considerate person of his age I've ever met. So we do well, each going his curious way. Jon bought him a horn and he's becoming a swing enthusiast of consequence. The girls are with their mother; now in Puget Sd.; they just returned from Sitka. The old crowd has changed. Francis and Elaine Whittaker alone remain. Frank Lloyd was in last night for a few minutes. Hilary [Belloc] is at war. Red Black thinks he also may go. The war is imminent. Or something is imminent. The old easy life that we knew (not financially easy, I mean, but intellectually so) is I think going, not ever to return in our lives. The monasteries kept a torch of intellectual activity if not freedom during the centuries of apparent patternlessness. If we are likely to descend into another dark age, I wonder what'll keep it alight now. I often used to think to Nan— mostly in my own mind "We are getting along; life is short; let's keep something out of it." Like Matthew Arnold's "Dover Beach." So many people have had that thought, on varying levels of superficiality and significance. "Gather ye rosebuds" naïve and good. "Eat drink and be merry" in the more sophisticated but cynical vein. And Arnold's rather noble weltschmertz [world weariness]. Or

Wagner's words in Goethe's *Faust* "Ach Gott die Kunst ist lang, und kurz ist unser Leben" [Oh God the art is long and brief is our life]. Well in my middle age I wax philosophical. But I feel alright. No one can make a deeper contribution to life than by having fun, if in dancing, or writing, or going places, or making a good relation; so I say if that understanding can emerge eventually from a thing even so drastic as war, then I'm not against it. Are you getting shaggy dog stories there? I suppose that's where they come from. The horse who bit himself on the forehead; and the puzzled hearer who is reassured by the teller: "He had a terrible time doing it; had to get up on a stool." Or the wag in Hollywood who called up his movie director friend at 4 AM in New York by long distance "I can talk; isn't it wonderful." "Oh for christsake let me alone I'm sleepy." "But I'm talking; don't you hear me." "Well what of it." "But you don't understand, I'm a horse." Long letters to take up your time and mine; and we both no doubt have more urgent things.

∽

To Gwyn Conger (Steinbeck)
April 11, 1941

Toni typed off a slightly modified (in the first paragraph) transcription of this and we sent it, with ck. for $100.00 by [special delivery registered] mail. Mailed on depot platform at Salinas 9:45 PM Fri 11th April 1941

Dear Gwen:
This letter comes from me rather than Jon at his own suggestion. He is in the first place a very muddled man, knocked out with nembutal etc, and under very terrible strain, torn between what he wants to do and what he thinks he ought to do. Said he thought maybe I could express to you better how he felt than he could himself, in his present condition. And of course you won't relish any delay, you must be pretty much on tenderhooks. You know Gwen, I think you're standing up very well and very maturely before conditions that might easily sweep people under, and I admire you for it.

He seems to be sure of only a few things, foremost of which is the reality of his love for you and yours for him. The other I don't know if he's quite so sure about, but just now he thinks that he must stay with Carol at least for the time being. A large feature of this attitude is his compassion towards her as an inadequate unadult person who would be completely lost and dead without him. It's a great tribute to you Gwen dear, that John needn't have this worry in regard to you to torture him still further.

You will know by this that he recognizes that it is you who love him best, and that he loves you the best. So far thruout he isn't fooled about his feeling towards Carol. His good compassion is an important factor in making him want still to help her as much as he can. Which, I should say, can't be very much, because you can't give a person what you don't have, or what you don't have for him, however good your intentions. But he doesn't realize that yet, and saying it of course doesn't do any good.

It is essential that he be able to draw strength from knowing that you're safe and alright, and that you'll continue to be alright, however things fall eventually between you. At any rate he'll feel easier knowing that you can reconstruct your life, and that you aren't putting on him the pressures of immediacy and upset and illness that Carol seems to be impelled to do. I think that the gains achieved thru immediacies of this sort have no lasting significance with any one so fundamentally sound as I take Jon to be. The pressures of hysteria and suicidal inclinations that Carol seems to be putting on him now, consciously or unconsciously, complicate matters and make it impossible for him to come to any clear vision of the ultimate path of his own life. There's probably no point in my saying this to Carol, maybe still further her confidence, but it's true and may make you feel good to hear that in this thing so far you seem to me to have acted like a woman in love, while Carol seems to be acting like a threatened person. And that your actions seem to me first of all dictated by love, and only secondarily if at all by expediency. And that Crl seems to me to have acted first of all from the standpoint of expediency, and only secondarily from love. And I say this loving Carol plenty but not blindly.

Probably you'll feel good, as I do, to know that thruout Jon has stuck well by the one thing he has been sure of: the validity and reality of your love and his. He hasn't and won't, (even to Carol) belie in any way his belief in the depth and truth of that love. He has seen that sure light and stands up to it consistently.

This letter is written hastily, and in attempting to work out and say what seems to be true I may have shorn it of the many tendernesses I feel I could ordinarily express. I haven't had time even to make sure of the deep trueness of all these things, so you'll have to forgive me if I'm anywhere in error. But within these limitations, and working with whatever love and wisdom I can muster, I should say in summing up that the best move for you personally and as a friend of Jon is to reconstruct your life as tho nothing further were to be expected from that source, at least for an indefinite time, and to be sure of his love and loyalty towards you as you can be sure of mine.

The attached will get you to Chicago if it seems to you still wise to go there. If you do, don't forget to keep in touch. Have fun. And know that Jon doesn't think you're deserting him anymore than you think he deserts you.

⁓

To Nancy Jane Ricketts
June 3, 1941

Dear Nancy Jane:
Several things. I was away for awhile in the S. LA; looked up two or three of the specialists who are going over specimens we took in the gulf, did some library research etc. And yours and Cornelia's letters are up for pleasant answer, having been here probably now for [a] couple of weeks.

Ed is already to graduate. Thursday, this is Tuesday. He has asked Fred and Aunto and I, and Jon and Gwen (Jon and Crl are separated now and he is with Gwen whom you didn't know). They are very glad that he asked them and he was glad they wanted to come. Then he is driving Toby Street up to Lake Tahoe on the 13th.

Ethel and Earl just went back to LA. (Maybe you remember her. We stayed at her apartment one time in Los Angeles, and you and Bee kept going up and down in the "alligator" until the other people in the apartment, and Ethel's room-mate, were pretty nearly reduced to tears). Ethel had a baby, now eight months old. A very happy and lovable girl. Also Tal had a baby. Maybe I told you all this. And Barbara is going to. And they are going to Mexico first to paint, and were in this afternoon to get dope on prices, regulations etc. Also Marj Lloyd is going to have another baby. Everyone. Francis and Fred also were in this afternoon. Trying to get Ed some tweed trousers for graduation.

The book is going fine. But slow. It's going to be a wonderfully good job. I think will live a long time. But so long, and so much expense. We're having a lot of color photos made. Elwood did Jon's portrait, a fine job.

Isn't that good about the dancing. I'll bet you're a fine dancer. Soon maybe we'll be hearing you on the radio. Monteverdi and Vittoria—also spelled Victoria don't know which is right—and Palestrina, and Purcell and Gluck and Pergolesi and all the old writers of madrigals and motets before the different 19th century Solemnis. The lab is just about the same except that I have a lot bigger library. Just started a year or so ago to build up my scientific reprint library again. Ed has a car; oh I guess I told Cornelia that R and Tal gave him their old one. Monterey is much changed. An army town, I guess like Bremerton. Best love to you and the folks dear Nancy Jane butterfly name.

> Nancy Jane Ricketts was sixteen years old at this time. She turned seventeen on November 28, 1941.
> Cornelia Ricketts's nickname was "Bee."

To Cornelia Ricketts
June 3, 1941

Dear Cornelia:
Bless your dear heart for sending me that photo. No, I'm glad you didn't have it tinted. I like it just that way. Tonight Jon and Barbara and Elwood all were looking at it. They said that when they saw you last you were just a little thing and now look at you!

Fine about the school. All those good marks should make it much easier for you to go on to college if you want to. I agree with you about the music situation; the "book-learning" of it doesn't seem to me very important either. Ed is pretty good now with his trumpet, and he learned all that secondarily. Taught himself from listening to the radio and to records. Now he writes down his own music from listening to records, and then plays it back on his trumpet. He graduates in a couple of days as I mentioned in my letter to Nancy Jane. And then drives up to Tahoe.

Would be so nice to see you. I had hoped that maybe this summer I'd get up north. But the book isn't done yet. And that's the first thing. So long! Now I have been working on it most intensely for nearly a year and a half. Since before the first Mexican trip last year. So I guess again this year I won't have any johnnies to go around.

I still have the same car. Now almost up to 60,000 miles and getting pretty shabby. Oh by the way I saw de Laubenfels recently. Don't know if you remember him, but your mother will. They used to live here. His wife used to be interested in riding horses. He is still at Pasadena Jnr College and still is working on sponges. I wanted to see him about the sponges we collected last year down in the gulf.

Well have fun honey and don't study too hard. Love to you and to your mother.

    Cornelia Ricketts was fifteen years old at this time.

Henry Miller arrived in Los Angeles during the summer of 1941. He initially stayed with friends Man Ray and his wife, Juliet, and through them met Gilbert and Margaret Neiman. During his stay in California, he traveled throughout the state with various friends, often the Neimans. It was on one such trip that he met Ricketts and Steinbeck in Monterey. Although the details of their meeting are somewhat vague, Miller's immediate interest in Ricketts is apparent in their correspondence.

To Henry Miller
July 2, 1941

Dear Henry Miller:
I have a feeling that it was one of those good things when you came here—when you wrote Jon, when Toni that good nice Toni recognized you in the maze of Jon's mail that ordinarily goes into the waste basket, the whole pattern of things. (Incidentally if I may not have mentioned it: Jon also likes you which makes it unanimous and hopes with me that you can come back here soon). The people are few who speak from "the other side"—it was obvious to me that you must be such a one from what Toni said of your work and from the little I read of it, and as I knew soon after I met you. It's good if those few can know each other, and better if they get on well. I feel as tho I am enriched and as tho I on my part also contributed to an enrichment. So it's one of those wholly good things. Isn't this swank lovely paper. Merits better typing. We use it for first draft of the new book.

No hurry by the way about those essays. I thought they would be right down your alley, and I feel grateful to myself at having written them if even for no other reason. Would you be interested in having a copy of the scientific or rather semi-scientific "Btwn Pacific Tides?" Not at all in your line. Straight factual writing and not very wonderful writing at that. Now I feel diffident about having suggested it, but then fine alright. No harm, I will have one sent. It isn't meant to be read, tho; mainly a reference book.

Your two welcome letters at hand. Gave me a feeling of being well received, as your visit here must have given you. And the —— [?] thing, Toni is reading but I haven't touched yet. Went thru some of the "World of Sex" MS and am impelled to read it carefully. I am very grateful. Seems to me something of a personal credo with supra-personal even universal implications. (Curious that those beyond qualities shall ride on things that are traditionally taboo or terrible, Whitman's "death," Jeffers "night"). The sort of thing a person has to do in order to explain conceptually or symbolically the ideas he tries to put across orally or implicitly. It was that same quality of inquiry and mis-understanding on the part of friends that led me originally to work out those essays which also weren't originally intended for publication. Incidentally only one of them, the non-teleological thing—is scheduled for publication. This fall, in the new book (which will itself be something of a miscellany). I have no immediate plans for the others. Seems not the right time now. They are, as you say, too radical for immediate acceptance, at least from an unknown.—Altho the idea of their radicalness came to me originally as a shock, they seem just naturally "so," neither radical nor conservative. If the Sea of Cortez is well received, and if subsequent book or books go well (the SF Bay account is to be sort of a high school text, wotta stunt that'll be for the young idea; and then maybe one on the Aleutian trip if we take it) maybe it will fall right if I try for a book of essays including a couple more I have floating around in my mind.

I am glad to be able to say that Jn's first draft of the Sea of Cortez narrative was done yesterday, and he starts tomorrow on the second which is to be first typescript. It came out very very well. I only hope some people know what we're talking about. Not too many tho. If they get enraged at the desecration of idols. The business of calling things by their right names is dynamite. Or even by their approximately right names. "die Weingen . . . hat man von je gekreusight und verbrannt." Which needn't make a person deist, but can let him know what to expect—which is never so distressing as being taken unawares in simple confidence betrayed.

I have one of the Zen Buddhist books, first volume of the Susuki essays. The other things I hadn't known. Some of Fenollosa, maybe in connection with Ezra Pound's Li Po transcriptions. I understand Rudhyar used to live here, maybe still does. Toni's father told us a while back that we'd do well to know him. Oh yes and going back to your MS; Toni told me that your essays considered what I have been calling "breaking thru," and I find it again there. Now it's almost midnight again. The typewriter is going in the other room—this will only just now be quiet—Toni is finishing up a great long scientific section, very dry. I will turn this over to her in case she will add a word. Regards to Gilbert and Margaret. Don't let them drive back without stopping here at PG. We are hoping to see you soon.

Ed Ricketts

On June 26, 1941, Miller wrote to Ricketts thanking him for an intriguing visit, stating that his time in Monterey was a highlight during his travels in the United States. Miller also mentioned how much he enjoyed hearing the portions of Steinbeck's writing that Steinbeck read aloud during their evening together.

Ricketts sent Miller his three essays, and later in June 1941 Miller wrote back to Ricketts. "Breaking Through" was most interesting to Miller, who went on to say how much he admired the clarity in which Ricketts wrote about his ideas in all three essays. Miller also commented on *Between Pacific Tides,* again noting the accessible style of the book and Ricketts's writing.

To Henry Miller
July 1941

Henry Miller
1835 Camino Palermo
Hollywood, Calif.

Dear Henry:
If I write you at once, I shall have recorded your new address.

By all means come along up next week end. There'll be room either at the house or at the lab, or at both, unless someone else stops by, and it isn't likely that everything can possibly be taken up. Come on.

Essays received back OK. I'm very glad I had copies of them for you to look at. I would have been glad to hear from Gilbert about them. And would still.

Stanford were willing to print only 1000 copies of the invertebrate book. That was 2 years back. Before fall is over a new printing will be needed. And I think we'll make it a new edition, altho where I can get the time to work it up is hard to see. As a financial proposition it has been (for me—don't know how publishers made out) a perfect flop. As I anticipated. In order to insure the sort of illustrations we wanted, I had to make royalty concessions. No income at all from first 500. My net receipts have been $30 so far. But I had to send out a great many copies to collaborators. Now that's pretty well over, and I expect to get in $100 or more per year from it for the rest of my life. This one will be better, but probably not so steady. So

Ed

Miller hoped to visit Ricketts and Toni near the end of July 1941 but was unable.

To John Steinbeck
August 22, 1941

Information has come in now sufficiently so that I am making general summaries, and I think you will be interested. It would be an understatement for me to say that this little trip of ours is proving to be an important expedition, and that out of it are coming some fairly significant contributions to invertebrate zoology, to marine sociology, and even—I wouldn't be surprised—to human thought.

There are several things about the trip, even if regarded as an expedition, that strike me as being fairly unique. With a fishing boat not specially equipped, with time limited to six weeks—one third of it spent in getting there and back, without laboratory facilities, and, finally, with personnel limited to two or three workers, we seem to have got the rather considerable total of some 600 species of which up to 50 may be new, and many others of which are proving to be rediscoveries of animals reported only once and hence regarded as rare (altho they're common enough in their characteristic ecological niche). This is all the more remarkable when you consider that: (1) we worked solely in a small, but natural, subdivision of a single zoogeographical province; had we taken specimens along the way, from here on down there, as the average expedition, interested solely in numbers, would have done, we could have collected in three provinces, with a corresponding increase in the number of species taken; (2) we took shore animals solely—since we didn't even have any dredging equipment on board—and 95% or more of the material was strictly intertidal; and (3) we paid very little attention, except incidentally, to (a) such pelagic materials as jellyfish, ctenophores, micro-crustacea and other planktonic elements, (b) fishes, except for the commonest shore species which were important elements in the sociology of the region, and (e) rarities, small animals, or obscure forms.

The average expedition takes material indiscriminately by dredging, seining, tow-netting both in deep water and in the surface and usually far from land, and pays especial attention to rare or unknown forms. We operated on the opposite plan: the commoner the animal the more attention we devoted to it, since it, more than the total of all rare forms, was important in the biological economy. Instead of operating on any hit and miss plan, we had merely the one coordinated objective: to become as familiar as we might, in our limited time, with the shore biology of the region—to make a survey of the common and obvious animals of a restricted area.

I think we not only succeeded, but that we succeeded beyond our hopes. If even we had no limited objective, and if we had worked without any geographical or ecological restrictions whatsoever, still we'd have done well to get such a bunch of material. And getting it all whipped into shape for a thorough report within less than two years is almost unprecedented. I think we have done a good job by any standards.

And furthermore, said he, we few, with no decent facilities, and with the laboratory, library and office difficulties incident to a small boat, and with no coordinated force of specialists and clerks, collected, prepared (usually well), sorted, labelled, dispatched and kept our records straight on this enormous bulk of material without any grave confusions, whereas many a pretentious expedition got back with lots of their specimens poorly preserved, unlabeled, mislabelled, or lacking field notes.

It seems gratifying to reflect on the fact that we, unsupported and unaided, seem to have taken more species, in greater number, and better preserved, than expeditions more pretentious and endowed, as we were not, with prestige, personnel, equipment and financial backing. As Toni says: "Two guys in a small boat, with enthusiasm and knowledge."

It appears that our unpretentious trip may have achieved results comparable to those of far more elaborate expeditions, and certainly more unified and ordered in an architectural sense. It may well prove to be, considering its limitations, one of the important expeditions of these times.

Part of this information is being abstracted for the biological summary, but I imagine you will be glad to have it now, and in this informal way.

Sincerely, and with very great dignity, I am,

To Pascal Covici
September 15, 1941

Mr. Pascal Covici,
Viking Press, Inc.
18 E. 48th St.,
New York City, N.Y.

Dear Pat:
I very much appreciate your thoughtfulness in sending me the poetry anthology; I am all the more appreciative because the subject becomes increasingly my major special interest (oh, I haven't yet forsaken the lovely fleshpots—I mean in a mental sense); most of all because the job seems to have been very well done. From what little I've seen of it so far, I'm inclined to rank it as one of the three great modern anthologies of verse, altho subserving a function as different from the other two as they do from each other: Van Doren's *World Anthology,* and the MacLeish *New Anthology of Modern Poetry.*

Many thanks to you.
                                        Sincerely

To Gilbert and Margaret Neiman
September 29, 1941

Dear Gilbert and Margaret:
I had intended for sometime replying to Gilbert's letter. But it had no address, and if there were one on the envelope it's gone now with the discarded envelope. Then I discovered a letter from Margaret to Toni, this time with the address. And so that's the way this is.

Toni, who at best is a most irregular correspondent, has been recently even busier than I. If that's possible. Which it isn't. Taking down Jn's and Milly's words of wisdom, pearls of wisdom, for the *Red Pony* which will be filmed soon. And so she didn't answer.

The last of "Sea of Cortez" M/S has been sent, and the first part of the galleys corrected. Still lots of work on the scientific galleys, and on illustrations. But the good end is in sight.

We had a note from Henry sometime back. Said Margaret was in the hospital or scheduled for it. Something like that. Back now, I suppose. The hospital situation is hard times. I have been thru it often with my family. My ex-family.

We have thought of you often lately. Especially I have been seeing, but I guess this was now some weeks back, someone I identified with Margaret, way back in my mind. Some people click. All of us did. This is a good place incidentally. Many good people, of whom you met only a fraction. No, that's right, there was a house warming. You met most of them but few dependable impressions can be got in one meeting. I am sorry Henry went back east, as I gather he finally did, without getting up here again. Don't you people also commit that sin of omission.

The non-teleological essay thing went into the book intact. Jon eliminated the personalities and made one or two word or phrase revisions for the better, but it's 99 34/100% as it was. Galley was turned over to the Sci. Editor of the *NY Times*, I think it was, who got behind it most heartily. Might even be a best seller. In which case we'll get back our expenses and advances.

The essays on poetry and on the breaking thru thing aren't in any immediate plans. Sometime I may be able to get out a volume including these, a reprint of the non-tel[eological] item, and some new things stirring around in the back of my head. Lots of things in the poetry essay probably could be (1) made clearer—for instance I didn't intend to class either Jeffers or Francis Thompson in the 4th category (2) revised as the clarified issues turn out any confusions that I may have got into. There is some evidence that two problems are intertwined, the "word of God made man" idea, and the morphological stages of the process. Sometime I may have a fling at it again (those things take so much time! I spent many hundreds of hours on each of those), but seems now as the two separate essays would have to be done, in order to cover the material justly. The first would approach the situation from the standpoint of content. This is what I did, so that's already done. The second would consider the non-objective situation, the abstract, the divine geometry, whatever name you use. I'm a little bit lost there, altho occasionally I get a real contact. Those two worlds are gulfs apart. Any liaison between them is rare or temporary; but any truly unified field hypothesis must consider both. Until a few years ago I didn't even appreciate the reality of this "other" side, the imagist etc situation (symbolized for me in painting by Mondrian). I knew it existed, and I must respect it, because people I respected were in it. But it existed the way the north pole exists, or some rarely seen trade or profession. A few years ago I was visiting with some people in SF who were hosts to John Cage and Xenia Kashevaroff. I think you knew or know them. He's a modern composer. Percussion music. Has an ensemble in which Xen participates. Friends of Russel etc, ex-student of Schonberg whose work he no longer admires. Anyway we had one of those interminable discussions in which nothing seems accomplished, you get a continuous sense of frustration out of which only one or two rays of light emerge (which however *may* illuminate the whole thing). I saw nothing in Mondrian, little enough in Klee, Kandinsky, Leger, Braque, etc, not much more in Javlensky (spell. doubtful) [Aleksei Yavlensky] or late Picasso. They saw nothing in the representationists, even in the later ones, and Renoir became the symbol of what they disliked (or misunderstood) just as Mondrian symbolized the unappreciated side for me. We were good honest people who had the courage of our convictions, who were talking more than for the sake of arguing, and we were good friends. Anyway—this was all very dramatic—the following morning Jon said he woke up to look across the room where a Renoir reproduction was hanging, and suddenly it all fell into place, he saw the planes that made up the composition, appreciated and approved. And since then I have had at least some contact with what they called "inspired geometry," the art of form rather than of content, in which so much modern poetry sits. All this is apropos in the sense that when John Cage and I were last discussing the poetry essay, and hence poetry, his definition, which I wish I had noted at the time, was just about the opposite of mine.

Well. Long letter. Started out pretty much just to extend greetings. Come see us when you can. Toni's clicking away on the other typewriter, says hello. I'd leave this letter on her desk for her to add a word or two, but if I did Lord only knows when you'd get it. Regards to you both, and I hope Margaret's better.

> Gilbert and Margaret Neiman were friends of Henry Miller who lived in southern California. Gilbert published a novel in 1947 about Mexico entitled *There Is a Tyrant in Every Country*.
>
> Steinbeck worked on the screenplay for *The Red Pony* with Lewis ("Milly") Milestone in September 1941, which was not released until early 1949 due to delays in filming. The two had previously collaborated on the film adaptation of *Of Mice and Men* and shared a mutual respect and friendship. Toni worked for Steinbeck typing sections of *Sea of Cortez* and also assisted with *The Red Pony*.

To Gwyn Conger (Steinbeck)
November 3 or 4, 1941

About Friday or Monday
3rd or 4th
Nov-1941

Dear Gwen:
I am terribly sorry you had all the grief of feeling bad (a grief I know very well from old times) when, actually, there was no factual and external reason for you to feel out in the cold. From any of us. When a person has to face an external difficulty, it's hard enough, altho necessary, to do so. In this case the facts were otherwise. Crl got terribly little. Too little. With Toni and with me especially, she was herself constrained, and I hadn't for her, at the moment, that

love that washes over all constraint. In addition, I was too busy, and saw not as much of her as I would want to see of a friend who felt, whatever the cause, bereft and low. Even if she got much, tho, it wouldn't take from what we have for you, would rather increase it. But I'm glad you wrote. If you're anything like I am, it's a good thing even to get an insecurity sufficiently objectified to put it down on paper.

Toni had been writing—and took a good deal of joy in doing so, as you will see from her letter—but was and is careless about finishing and mailing letters. I had my hands so full, I was putting off (as usual, wotta life) everything I could possibly put off. Tal doesn't ever write. I never knew her to. To me anyway. I suppose she can write. Of course, I've even known her to write her name. I don't mean she's illiterate. Or is she? And guys like Toby and Barb and Ellwood are just too plain inarticulate. Besides, they never learned to write. It was too hard. Like working a harmonic analyzer, the great brass brain, not many people can learn to use it.

Ellwood had a show at DM [Del Monte Hotel]. Still on for that matter. Toni wrote you about it. And you'll get the letter. Because I mailed it myself. That is I probably did, and I've forgot what I did with my coat so I can't check to see whether it may be still be in my pocket. No I'm sure it isn't. Practically sure anyway. Barbara has a show at SF, but we haven't seen it. Maybe Toni'll get to it tomorrow, while I'm at UC reading, this time for a last fling at the bibliographic addendum for reprinting of *Btwn Pac Tides*.

Haven't seen much of Toby. Went over to Beth and Peg's Sat and again Sunday. I'm still on the wagon so am pulling my punches. Not a drop. They had a most realistic old-time bar fixed up. And Tal and Ritch lent a hand in making a dummy that looked like a dead man shot over a game of cards. Really v.g. Ritch did a fine monologue on the death of some sow bugs, and the rest of the gang a script (improvised) on a woman's ambulance corp adventure. Ritch is working hard on two writing jobs now. Kay is at her grandma's. She had a swank wonderful party for 10 girls. Toni must have been worn out. I was, even from picking them up and delivering them. But every one had a grand time. Elwood just had all his teeth out. Got ten dollars apiece from several of us. Remembering how Jon gave me $100 to have mine out, I kicked in gladly, told them it was a gift, not a loan, and felt very good about it. Ellwood did some grand work on that Mexican trip. Since then tho he's been unable to work, and recently, unable to sleep. Had 8 abscesses. Maybe that malaria didn't do him any good. Somehow he got some pleurisy out of it. I imagine the lung equivalent of black water fever which old timers used to say was apt to follow a very severe case of malaria. Take your quinine now Gwennie like a good little girl.

I suppose Jon told you, and will have told you now, about the last lot of reproductions, I felt very bad about them and wrote him immediately. Wrote McIntosh and Otis also. And Viking. Hope some of it does some good. If Viking makes a good job of that book, it'll be over *their* dead body. Yet they can get out good books, even from standpoint of format. That new anthology of English verse seems to be a fine job. I just got for Toni "Black Lamb and Gray Falcon," the new 2 vol. Rebecca West acct of Jugoslavia which they got out. Altho not so striking from standpoint of physical book-making, I think (in which I agree with Fadiman which of course makes him a very great critic!) it's a very great book. The great book of this year. Strikingly similar to what we tried to do for the Gulf, I'd say. Toni very thrilled. I picked it up at the Village Book Shop in Crml after had tried unsuccessfully to get it at Miss Smith's.

So, coming to the end of the paper. This is always why I stop writing letters. You can't win. The end of the paper always pops up. Gwen honey, feel happy. At least feel happy about outside things, the other is everyone's problem. I think this is your home. And don't worry. I don't think anything's going to hurt it.

> Ellwood Graham met and befriended Steinbeck and Ricketts in the early 1930s when Graham was painting murals for the Works Progress Administration, and Steinbeck took an active interest in Graham's work. In 1941, Steinbeck commissioned Graham to paint his

portrait while he worked on *Sea of Cortez*. At the same time, his wife, Barbara Stevenson—who later exhibited under her given name, Judith Deim—painted Steinbeck at his writing desk. When they finished their portraits, the Grahams traveled to Mexico with financial backing from Steinbeck.

Ricketts and Steinbeck were frustrated with the publication logistics for *Sea of Cortez*. As Steinbeck biographer Jackson J. Benson notes, "The Viking editors seemed suspicious of the collaboration and reluctant to invest in the illustrations and special typefaces required for the catalogue, so there was much correspondence back and forth during the summer of 1941 about these and other details of publication" (482).

⌒

To Joel W. Hedgpeth
November 3, 1941

Dear Joel:
It may seem strange to you that I am answering now a letter I received from you last December. Or maybe it won't seem so unbelievably curious. Zoologists are a funny lot.

The *Sea of Cortez* is done, gone, off finally to the publishers, with proofs corrected, index made, even the illustration dummies checked. So that now I can take forty winks to get caught up.

It has been decided to make the new printing of *Btwn Pac Tides* not a complete new edition, but a revised reprinting in which we plan not to open the forms, but to make a few addenda. One of them involves the literature of the past 2 or 3 years, and I am citing your key with a remark that its bibliography includes everything apropos (on this coast) since 1938, with mention specifically of the two pretty jinglebollix papers (streamlined spelling!) and a couple of yours. I didn't even look up the Hilton opera. A pity that he persists, and a greater pity that the Pomona [Journal] goes on. The home of mis-spelling and improper citations! Seems to have been infectious. It isn't only Hilton that has made mistakes in it, everyone who writes for it does! I wonder if the PG new genus [*pycnogonid*] has anything to do with some stuff I sent him from here 4 or 5 years back. I'd hate to think I had anything to do with perpetrating the errors he'll sprout.

Jon is in NY, writing a play. Don't know when he'll be back. We have been getting your papers right along, and gratefully. The F and G Divn thing just in.

No use talking about last winter's Christmas tides now, but if you're around here this Christmas, I may go out. Will do a little work on the Tomales Bay and SF Bay mudflats sometime soon, maybe then. But will probably be here.

Regards,
Ed

> Joel W. Hedgpeth (1911–), a marine biologist, met Ricketts "through correspondence about the pycnogonids to be included in *Between Pacific Tides* in 1935" (Hedgpeth, "Philosophy" 101). In late 1938 or early 1939 they met in person; they saw each other for the last time in 1945. They continued their correspondence until Ricketts's death, and Hedgpeth subsequently prepared later editions of *Between Pacific Tides*. He also edited the two-volume collection of Ricketts's essays, *The Outer Shores* (1978), and published numerous articles about Ricketts.

After separating from Carol, Steinbeck moved to New York with Gwyn. Carol had been living in New York for a short period, and upon her return to Monterey in October 1941 she became somewhat aggressive due to frustration over Steinbeck's relationship with Gwyn. Steinbeck hired his friend Toby Street as his lawyer and asked him to work out a settlement with Carol. Ricketts found himself in the awkward position of mediator between Steinbeck and Carol.

To John Steinbeck
Pacific Biological Laboratories
Pacific Grove, California
November 7, 1941

Dear Jn:
Instead of there being no news, there is too much, and bad.

I gave you a bum steer on Carol. She is definitely out to devil you, or rather, Gwen, and where it hurts most,—in her sense of security. Mavis was entirely right. All of us were wrong. Instead of being OK and fine and getting along alright, apparently she is just as vindictive as ever, but just so quiet we didn't realize it.

When she left here last, she was angling for a SF apt., but worrying about the cost ($75 per month!!) Yesterday Toby called up about the rabbits, and later drove out and confirmed the fact that she is coming down here to live (after all the years' nonsense about being unable to adjust to the PG climate, and requiring Santa Clara County where she was raised). And plans to live in yours and Gwen's house, very possibly the only home Gwen ever had (which is how Crl can strike at Gwen's security—if Gwen feels about those things as most girls seem to feel—by living in and maybe trying to acquire that home). Maybe an intuitive pre-visioning of this was behind the very sad and insecure letter Gwen wrote to me and Toni last week.

I was more or less nominated, partly nominated myself, as a friend of Crl's, to tell Toby that how I felt about it was just a drop in the bucket to the way Tal and Ritch, Barb and Elwood, Beth and Peg felt about it, and that, as her attorney, she ought to know beforehand what a hornet's nest she's getting into. But Toby said he knew that, and she ought to get all the Hell that's coming to her on that score, and he thot I also would do well to let you know the feeling here about it. Toby is on the whole I think very clear and good. He says he'd rather see Crl attempt her worst right now and get it out of her system, instead of striking at you, or trying to live in your home, later.

I didn't know what to suggest about the rabbits except that I can't take the time to look after them up at our house, and we have no suitable place. Because you can count on one thing (whether or not those rabbits are a symbol to Gwen of her home): If Crl lives there something is going to happen to them. Maybe not intentional. Or maybe so, and only slightly masked.

But it's the effect on the good feeling of the house that bothers me. Worst of all the effect on Gwen's already not too robust sense of security. If that can be prevented, I suppose it won't be so bad. But you can tell Gwen, if you care to, that we're all with her, and for her not to worry. The Crl machinations are failing on fertile ground, and merely rebounding, more's the pity, to her own detriment.

When I first heard the news, I can't tell how low I felt. And frustrated. I was all mixed up and couldn't work it out. All I could be sure of was that I felt nasty and unfriendly towards Carol, and that bothered me too. Because I'm [a] friend of hers, and want to show it even more than usual now that she's (as it turned out) tending to alienate people by making such a vindictive ass of herself. Then I also felt terribly alone. I realized I should, and inevitably would, express to her my complete disapproval, and, not knowing then how the others felt in the matter, supposed I would be unsupported and alone. And I hadn't worked out yet how I could express my disapprobation of her *action* without disapproving of *her*.

And then I got to thinking I must be over sensitive (as, until that last lot of reproductions came in, I had got to assuming I must be hypercritical in the matter of *Sea of Cortez* illustrations). But when I told Toni she really went to town. Said that if Crl could do a bitchy thing like that, she wouldn't even stay in the same room with her. Wanted to get Barbara's reaction, and Tal's, and Beth-Peg's. I told her she mustn't even discuss the subject in so biased a manner, lest she indoctrinate friends of both and put ideas in their heads which weren't really their own.

But now I discover that all the rest feel as violently about it as Toni. Women have a sense of the importance of a home to another woman, and the girls especially seem to think that it's a basically vindictive act on Crl's part, intended to violate even, which they can't countenance without violating their own loyalty to Gwen. I'm pretty certain that if it ever comes to the point where loyalties toward Crl and Gwen are mutually exclusive Crl must inevitably lose. Of course people get over those things and finally get to countenancing even the greatest injustices if they are quietly and persistently enough continued. But in this case, I don't think anyone is going to go along with Carol's actions, tho we may if we're up to it go along with Crl herself. I'll be surprised if any of us step foot in that house while she's there.

But the girls especially fear that Gwen will be violated. Don't let her be, because we love her and love you, and if you folks do make a life together, it will be with the best wishes of all of us.

I guess everyone felt particularly shocked because we all felt that she was not only snapping out of this, but getting a deeper contact with herself. And that still may be so, and she may be really getting on her own feet finally, but that at first those new found feet are wobbly, and don't take her good places, or well. Still, it's a shock to discover she's still harboring designs apparently vindictive, and that merely by cleverness, she has misled us so far. With a person like that it's hard to separate falsehood from real truth. I don't mean the funny story or the exaggeration thing. But basic untruth maybe masquerading on more factually true statements than even a truthful would require. Her picture, as it would appear from her discussions awhile back, is that the suggestions for her living in the Eardley St. house came originally from Toby, and that it is by his indoctrination, against her judgment, that she would come to decide (if she ever did, and she seems to have decided) to live there. But Toby seems to be basically as sound as Carol is unsound, and the whole arrangement seems to have derived from her wishes, and through her ridiculously transparent rationalizings. She specifically told Peg that Toby had stated she must establish a 3 month's residence in this county, and Toby says, and I believe him and Tal confirms it, that he told her categorically the opposite, and emphasized it, since she seemed to be wanting a loophole to believe that. Also she told Peg, or so I understand about 4th hand from Toni, that she was merely awaiting from you the go ahead signal to get a divorce.

I think the answer is that Crl, whether or not she's getting better, is and probably always has been a little off her head. May even develop a persecution complex, since she'll probably go ahead even tho everybody's against her. And that she's at heart probably always has been, look at how angry she's been at Idell most of her life—pretty vindictive. And is able to convince herself, difficultly, that hers is a holy cause. To save you from yourself. And to save your fortune (altho she's dissipating it a lot faster keeping you away from where you'd rather be. Incidentally, she told Toby that she wanted to live in the Eardley St. house as an economy measure! That she regarded the thousand a month as coming from the principal, therefore fearing the indefinite continuance of it, and that therefore she had to economize furiously!). But underneath it, I think she's out to hurt you, Gwen particularly, by any means fair or foul that she can get her hands on. And she's clever. And she's going to hurt you, or Gwen, if any openings are left thru which she can do so. And all under the guise of casualness and friendliness which apparently fools, in the long run, no one but herself.

I don't know how you feel about the whole thing at the moment, and will go only by my own ideas from this end, and by my own feelings which certainly got wrenched. So will probably merely express my disapproval of her activities to her, and about her.

Anyway, I've tried to correct the very obviously wrong impression that I gave you in previous letters, and to clarify my own personal reactions. If you've left any openings by which she can hurt you or Gwen, she's sure as hell going to take full advantage of them; just like Mavis said.

Ed

~

To Joel W. Hedgpeth
November 18, 1941

Dear Joel:
Glad to hear you worked out a dicker with Joe. I am enclosing carbon of author's abstract; you needn't bother to return it, I have the original and we'll probably have to retype it anyway.

However much it seems otherwise "Sea of Cortez" is truly a compilation. Jon worked at the collecting and sorting of animals, and looked over some [of] the literature, including the specialist literature, and I had a hand even in the narrative, altho the planning and architecture of the first part of course is entirely his, as the planning of the scientific section is entirely mine. But much of the detail of the narrative is based on a journal I kept during the trip, and some of the text derives from it and from unpublished essays of mine. I shall be interested to see what the critics have to say about the various parts of the job in connection with the oft repeated assumption on their part that they can spot a person's writings anywhere by characteristic tricks of style and thinking.

However, all the above isn't for public use but merely for personal orientation. The book isn't so much a question of first half by Steinbeck, second half by Ricketts, as it is a true compilation in every sense of the word.

Also, there is a map, a good one. Or rather two of them. I feel just as you do about the map situation. The only trouble is that a good one takes so much time and energy, and a person has to put down a line of demarkation somewhere.

For the forthcoming reprinting of BPTides, no map is projected. (But one of the projects for the completely revised edition, on which I hope to start when present commitments are done, is a good map). By the way if you don't think it's something of a job or more than an evening or two, I'll make a little proposition: if you'd like to make up such a map, and can turn it in within the next two or three weeks, I'll be glad to kick in ten bucks towards it, and see that you get full credit for the work. I've even been thinking about doing it myself, but I'm terribly dumb and awkward about such things. It really ought to be done. I agree with you there. Just a problem of how and when. Under the above suggested circumstances, we could probably run it in the reprinted edition which should be rolling soon.

I will change your address as suggested. This can be done very readily in an errata list I hope they will publish. Also I did not pick up the p. 306 error you mention, and am glad to hear about it. I have a list of such corrections and will include this one. If you find any other downright errors—there are still plenty of them I have no doubt, despite the multiple combing the book has had so far—please let me know.

I suppose that Stanford will send a review copy to the *Chronicle* as before. If I recall correctly, Jon sent a copy to Joe two years ago, but I forgot if anything came of it. incidentally, SU Press is a pretty slow outfit. I haven't yet turned in my copy for the corrections to them, altho it's already (new royalty arrangements are being worked out *I hope*), but even after that it may take them a long time. It was nearly three years between the signing of the contract and actual publication in 1939.

I have received no duplicate galleys of the scientific appendix, and no page proof whatsoever. They were in such a hurry to put things thru that they apparently corrected the page proof themselves. However we went over the galleys with a fine tooth comb, and since they have one of the most proficient proof-reading departments I've ever seen, I think they will pick up all the errors to which their attention has been called.

[Note: The following typed paragraph was crossed out and followed by a handwritten note at the bottom of the page.]

If the MS will serve, I can send that to you—you saw part of it when you were here the other day and can see that it will be hard to work with. It's furthermore uncorrected. But if you want to supplement your hasty examination of the finished product with looking over the MS carbons, I think it can be arranged.

Let me know soon if I should send it.

No—on second thought *very* impractical since that would leave me with no way of answering any of the last minute telegraphed inquiries that have been coming in, and with no record.

༼

Sea of Cortez was published in the first week of December 1941, but the bombing of Pearl Harbor on December 7 drew attention away from its release. Reviews for the book were mostly favorable, although on December 6, 1941, Clifton Fadiman of the *New Yorker* wrote, "The title page says it was written by Mr. Steinbeck and Mr. Ricketts, but I think we may safely assume that Mr. Ricketts contributed some of the biology and Mr. Steinbeck all of the prose. [. . .] The scientific supplement of the book [. . .] very forbidding and scientific, and calculated to increase our awe of Mr. Steinbeck."

To Jewel Stevens
December 15, 1941

Dear Mr. Stevens:
I am very grateful to you for sending that review. First for the thing itself—since I wanted slightly at least to keep track of the book's reception there in the middle west (John sends me the NY reviews)—but mostly for the thoughtfulness in your sending it to me.

So far I have seen maybe 10 reviews, 2 on the Pacific coast, the others from the east. Yours was the only middle west report so far.

All are favorable except for that in the *SF Chronicle* which simply blasted the first part (which is supposed to be Jn's entirely) and which approved the second. Reviewed by different hands. I would suppose that, as the book gets better known, it will become more controversial, and probably bitterly so. I personally rate it rather highly. But of course!

Only Charles Poore of the *NY Times* discerned that it was truly a collaboration. The essay on non-teleological thinking is mine and excerpts here and there are from other unpublished essays of mine. Incidentally—this is amusing—I tried to get *Harper's* or *Atlantic* to publish that essay and they wouldn't hear of it. But in a collaboration with Steinbeck, its publication offers little difficulty. The idea and the work and the words on the mutation of projected material, which I consider the most important contribution of its sort in the book, is Jn's. However, part of the fun is to see how and what the reviewers guess. I hope among other things that the book will sell well enough so that at least I can take care of the debts I incurred in the writing of it. The arrangement is that I am to pay back to Jn my share of the expenses of chartering that boat if the royalties justify it. Which they very easily could. Anyway I wanted to do some work down there and to get a picture of the region. And we had fun doing it. I spent the following summer in Mexico City working on it.

Mother is constantly getting better. For awhile, before that last attack, she was peppier than she had been in years. You probably heard that Thayer was out. Next time Pat comes out here we hope to see you too. A vacation never hurts a person.

Ed is working and saving his money. Presumably for college expenses. But he'll probably go to war instead. As who won't! The second time for me unless my false teeth and bum stomach keep me out. Which probably they will.

Well. Again thank you. And hello to Pat.

Ed

Jewel and Pat Stevens, who lived near Chicago, were friends of Ricketts's sister, Frances Strong.

To Joseph Campbell
December 31, 1941

Dear Joe:
Glad to hear from you. Even tho, to get such a response, I had to write a book!

You got the gist of what was intended, not only better than anyone else, but as no one else. Not one of the critics to date knew, or at least stated their knowledge, of what it was all about. All of them however liked it (except old sour puss Joe Jackson [San Francisco *Chronicle* book critic], whose man Friday regards philosophical speculations as confusing!), and several quoted very significant parts, still without getting a sense of the whole.

I was very charmed with the book. Jn certainly built it carefully. The increasing hints towards purity of thinking, then building up toward the center of the book, on Easter Sunday, with the non-tel essay. The little waves at the start and the little waves at the finish, and the working out of the microcosm-macrocosm thing towards the end. I read it over more than I do lots of other things still. Well, it's nice to like something you have a hand in. I figured you'd like it too. Right down your alley. And doubly so because you had a hand in some of the ideas and collecting details both.

Incidentally, Henry Miller was here for a while. I think he's on that same track, more fiercely than clearly, but not unclearly. Some of it comes out pretty plainly in the most recent thing "The Collossus of Maurousi." I think he's a good guy. He also read, apparently liked very much, and praised very highly those three essays of mine. You may have run onto him; last I heard he was headed for the east.

I'll look forward to seeing your work on *Finnegans Wake* and on mythology. It really ought to be done. Only trouble, there again, is the matter of erudition. Both on the part of the writer and reader. I have barely glanced at the book, and know *Ulysses* not better, altho I have the latter in the Mod. Libr. Edition. In order to see the whole picture, a person has to put so darn much time into it! And human energy so limited! This is the way I seem always to end a letter. End of the page!

Best wishes to both of you

Ed

To Nan Ricketts
March 13, 1942

Dear Nan:
A good letter. Made me feel very warm. I answered it at once because I thought it might inquire about data you needed for income tax report, and I'd tend to it at once. Tell the girls I will write first thing [when] I come back—going to Bkly this afternoon.

Your income from me, in case you have to have the figures, was $187.00. A miserable figure. Looks like no better now. I hate to see you work that way, but that seems to be the only way of doing things.

Royalties on the book [*Sea of Cortez*] have so far been nil, and again I am afraid I fathered a financial flop. Jn and I between us drew $2300 royalties, all of which and more we spent in photographs, drawings, steno fees, etc, to say nothing of the expense of the trip, for which I owe him half. But we did a good book, and we had fun doing it.

Nice funny Nan—wotta time to get up, 5 AM. Altho I do it sometimes myself, collecting etc.

Yes, you said it, this is a turning point in the world. In our western world anyway. Gradually things will be more and more different, and I doubt if they will ever go back.

Priorities have had such queer effects already. If I can't get tires, for instance, eventually I'll have to stop collecting, if chemical priorities haven't slowed me up before that. I got two more barrels of formalin thru Bogard, but alcohol simply can't be had. I got 5 gallons on the last trip, and I hope for another 5 today, but it takes actual personal contact. In order to get glycerine, you have to sign an affidavit that you'll have to close up shop without it. Which means no more embalmed cats. And I can't honestly say I have to have it, or them to get going.

Ed Jr has left and is working in the SF Bay area for the Calif. Packing Co., same outfit he has been working for, off and on, for the past six months. They told him if he'd agree not to go back to school—to university I mean, he graduated from [high school]—they'd put him on steadily up to the time he'd have to go to the army. He says they'll pay him not less than $32 per week. I wish I could earn that steadily.

Everything is very much different here. Sentries have been patrolling Cannery Row, but just discontinued. They'd challenge and shoot at someone almost every night. R and T have moved away. Carol got her divorce yesterday. I was witness for her. Jon is still in NY. His new book will be a best seller, will earn him lots of money. He gave Carol, in their financial settlement, all the money they had, except for income tax and $7,000 held out for him to live on. But now with this new book, he'll bounce right up. Sea of Cortez cost him a lot of money.

Keep well dear Nan as you can. Have a happy birthday, and love to the kids.

To Nancy Jane Ricketts
March 16, 1942

Dear Nancy Jane:
Letters from both you and Cornelia. Lucky daddy lucky daddy, ve like dem. And I have one other of your letters somewhere around here. Not right now reachable. Under a great mass of papers somewhere probably. Well, when I find it, it will be good reason for me to write another letter. And that'll be fine. But I probably won't.

Ed Jr, as I wrote to Cornelia, is away. Possibly for good. He was working in the office of Cal Pack here, mostly as assistant paymaster apparently, and then, when the canneries closed, they seem to have liked him so well that they suggested he work for them permanently—or at least until he's taken for the army—if he'd not go back to school. So he went to the head office of Cal Pack in SF, to sign on officially, and—as I think—to be broken into the way they want things done. I figure they'll keep him, right there until he is thoroly trained in their ways of doing things, but he expected, when he left here, to be sent right out to Antioch, in the asparagus canning country, or maybe to Sacramento. Anyway, he'll get a good salary, and probably have a lot of fun out on his own. He sold his car—the old jalopy that Ritchie gave him—for twelve dollars. It wouldn't run any more, apparently had a bum differential (altho we succeeded in getting it out to the junk yard under its own power, I drove and he followed along behind me in my car). So I have a bunch of his clothes, his radio, a selected lot of his swing records (he has over 300 now), etc, to take up to him when I go up in the bay region. But he isn't a very good correspondent, and I don't even know where he's to be permanently stationed. Probably he himself doesn't yet know. If he did he'd have at least his horn sent on. He can't get along without that.

I have still the same car. Now 5 years old, and never very wonderful at that. A pity I couldn't have got a new one before the priorities went in, but with care, I can get by. I have to be careful of money, more careful than ever before. Most people, especially you people, are earning more money than ever before, but I am earning less, and I think it's a chronic condition with me, and don't worry about it very much any more. I do some significant work, there are 2 good books that I largely had a hand in, and I do a lot of other (I think) significant things thru influencing and catalyzing. Or so I think. But for bringing home the bacon I never was very hot, and I'm getting less so. Which is alright too. There are lots of different kinds of work to be done in the world, and some of it is thankless and profitless, but nevertheless to be done. Remo—you probably remember him—is doing work of that same profitless sort now. Helping out Italians who are in trouble or under suspicion; trying to determine whether they are vicious or intentional, or if they've been led into difficulties by high pressuring from the Italian consuls, or friends, still in Italy, etc, etc.

Jon and Carol, as you probably read, finally split up. And good thing I say. They never did get along very well. Now Jon is with a girl he likes and who is good to him, and they get along well and have fun together. Crl and Jn had fun mostly apart. The chemistry is right between them, which is important altho maybe not all important, and they respect each others position. Gwen doesn't try to be the man and do the business arranging, bossing of the outside relations, etc, so that Jn can develop his own masculine tendencies which under Carol never were brought out. Jn could never tell what he was going to do, probably didn't even know, without having a cue from Carol, and Carol was sometimes a bit power driven. Now Jn can say what *they're* going to do without even consulting Gwen, because he knows in advance that Gwen, liking him, will approve. So he can feel free, and be relaxed and easy. A good thing, say I.

Probably you've read the new book "The Moon is Down." If you haven't, do so. It's a fine job. You can read it in a couple of hours. It is furthermore an important contribution to the United Nations war effort. I think he's done more out of the army than he could have done in it, because in military or naval life he would be subjected to constant supervision, and maybe some of his good ideas would be killed before they were born. (saving paper, using both sides). Have been having plenty of trouble with certain chemicals, and I suppose if things go much worse, PBL will be forced out of business. Bogard told me last week that he feared that within a year or so, nearly all small businesses would have to close shop. I'd hate to have to change now at my age, doing something else in place of fooling with the animals I've always liked. But I suppose those things happen too. Would make for a great capital loss too, because the lab building and grounds would probably stand idle, or have to be sold for some other purpose.

Cannery Row is certainly different. For a while we were under sort of martial law, and sentries challenged whoever went or came. One night they got after Tal when she was walking downstairs with their new baby. But now for some reason they have left and we are free as ever. There has been very little activity here since the war anyway. My personal opinion is that many of the fishermen are just plain scared. That's certainly one factor, but others may be involved. Some axis agents may have been working on their fears also.

Honey, did you graduate? Why otherwise are you out of school? I thought you were to get thru in June. That Bee has a great time doesn't she. First a mild case of mumps and now tonsillitis. Maybe if you get to working on your own, you can take a vacation in California. The land of sunshine—but not now. You should see it rain. And hail. If you could come down here I would like that. I have hopes to be in the Pug Sd region for collecting Gonionemus a tide or two this summer, but have to work out several things first. Finances mostly. But possible permissions too. I fear most of the shore line will be closed off. If only I had been collecting assiduously for the past two years, both here and away—in the north and south—it wouldn't be so bad. But I have been working on that book for nearly two years, and during that time also on another that Jon and I may sometime do—really a beauty. And I'm just about out of everything. Haven't even been getting many starfish. I hope to get down south next month, but have lots of preparing there to do. Think I can collect such things as the sea pansy Renilla at Corona del Mar, where MacGinitie heads the Calif Inst of Technology Station, but I don't know about the rocky outer coast of Lower California, etc.

You know if either you or Cornelia should ever be having an opportunity for a vacation we can put you up here in Ed Jr's bed. And it's pretty comfortable. He fixed himself a really compact cabin-like place.

Well there was something else I had in mind writing, but I've forgot it. Me and my fine fuzzy mind! So. Write when you can. I'm so glad to hear. Best love to you Nancy Jane my friend. Wotta long letter!

⌒

To Norman Carlson
March 25, 1942

Dear Norm:
Toni suggested that you might be interested in or might know someone who could profit by, some incidental information I ran onto today while chasing down references on the marine biology of the Pacific, in Hopkins Marine Station library.

I found Vols. (or Bulletins) No. 2 and 3 of a series got onto in Japan (but in English, as are most of the Japanese advanced biological papers), from the Palou Marine Biological station—name not certain, I can check it up easily enough. Palou (or Palao) also spelled Pelew is a Jap owned island, probably of strategic importance, between the Philippines and the Caroline Ids., where the Jap govt. established a marine biol. station sometime around 1935 or 36.

Some of the papers in this series that I hastily glanced over had very extensive data on the coral reefs, and of course on the marine animals of the area. I noted that one of the photos that was included merely to show a coral reef actually showed quite a bit of the mainland, with a huge radio tower. The first paper, wherein the station and its environs was described, undoubtedly would have extensive maps and photos, and probably charts to show depths, maybe currents, etc. Ecological surveys of a given region are usually very complete, the topography both submerged and above sea level is very important, and there's usually dope on water and air temperatures, rainfall, barometric pressures, etc., in addition to the information on the animals and their communities.

It occurred to us that, not only the above mentioned series, but other papers wherein the Japs describe the animals of their mandated colonies and of their shores (they are prolific workers and they publish most of this stuff in English, French, or German), might have information of potential value to US forces in connection with raids or military campaigns.

> Norman Carlson was in the army intelligence, stationed in San Francisco. He was a friend of Toni and corresponded periodically with her and Ricketts.
> Ricketts's interest in the Japanese mandated islands resulted in a significant collection of data and research, including maps and biological information, that he summarized and turned over to U.S. Navy intelligence. Steinbeck shared Ricketts's interest in seeing the U.S. government use the pertinent information available and wrote to the secretary of the navy regarding the matter. Ricketts continued his research during the next few years, hoping to eventually publish a book about the mandated islands. Due to a lack of enthusiasm by both the navy and various publishers Ricketts contacted, however, he eventually gave up on the project in 1944. On November 15, 1944, he wrote to Edward F. Ricketts Jr., "I gave up work on the Mandated Islands book. Figured I'd clear it up only if I could get contract for publication and none has been forthcoming."

To John Steinbeck
April 28, 1942

Jn:

You will recall that sometime back I wrote you, and you replied saying that the info had been frwd to the Asst Scry of War, on the info available in reports of the marine biol. of the Jap Mandated Islands.

I just took Ed with his horn, radio, records, etc, to his new timekeeping job in the sticks (Rio Vista) and en route back I stopped again at UC to have a look at this and other matters. I found that the info I had imagined might be there *was* there, and more. Those sweet Jap scientists, as pure and undiverted as scientists anywhere else, have made very careful examinations of the resources put before them in the mandated islands by their governments, and have reported fully. Incidentally including a good many maps (a Jap marine biologist helped the Admiralty Captain compile a new chart of Palao HQ [headquarters]) and much oblique data—oblique and interesting merely to a marine biologist, vital to raids and to war making.

Assuming that in this, as in other democratic procedures, the indictment "too little and too late" may apply in matters of strategic intelligence, I am sending copies of the abstract of this info to you there in NY, and to Norm [Carlson] in Army Intelligence here in SF. Wish I knew someone in Naval Intelligence where this would be most useful of all. (Unless already on hand there, which I doubt, this info necessarily *must* be part of the intelligence equipment of any planes or ships preparing to raid Palao, which is or was headquarters for all of the mandated islands). Maybe spreading it around in 4 or 5 different places, one of them might kick thru in time for the data to be of some use. If Naval Intelligence had conveyed graphically enough to Pearl Harbor what must have been happening in the mandated islands—info which I'll bet a cookie is inferentially available in Japanese marine biological reports accessible in this country—a better watch might have been maintained. Even tho the attack apparently didn't come from land.

Some of the oblique inferences in these Japanese biological reports are interesting. I am fairly certain for instance, that important powers in Japan enlisted the unconscious aid of Jap biologists in the working out of info, especially on the coral reefs, for improved marine charts. It's significant that the Jap gov't allowed travelling and living expenses to their scientists *only* who were willing to investigate coral reef problems. I'll bet that some very significant info was dug up in this way, that the Imperial Navy is making use of right now for clever bases in the Marshall and Caroline Islands. Funny part of it is that for all their secrecy in prohibiting white people from so much as sailing thru the islands, they shouldn't have realized that the guileless scientists whose aid they enlisted would inevitably not only publish the results of those careful studies, but would send copies of the papers to friends or co-workers all over the world. So here is the Jap nation now, potentially hoisted on its own petard of measuring and photographing and mapmaking and reporting. Now for the first time I have seen the value of political commissars, those much hated anhängen of all the "pure" enterprises in Russia. Well, it's all good clean fun said he recalling lovely women in the full of the moon. Well, hell, "k" really is right next to "l." On this typewriter anyway. And the [male/female] signs next to that. Wotta life! Can't you think of anything else. Haven't you any other purities at all.

And another thing. The examination of Japanese tropical medicine publications accessible in this country will turn out a lot of useful info on the mandated islands. Oblique info. Out of it all, a very accurate picture could be constructed out of a land otherwise unknown. Which we're going to have to deal with militarily or it'll deal with us. I'll look at some of these things as I get time. The search last week was particularly worthwhile, since I picked up the answers to some of the questions arising from the gulf trip. The careful Japs for instance have worked out the sea cucumber population of Palao sand flats, .047 to .565 per square meter, and the amount of sand ingested per year by the 407,400 inhabitants of 2 species on a sand flat of 800,000 sq [meters] at 12,460 tons. 15.60 kilos per sq. [meter]. Those cucumbers really do turn the sand over. Like earthworms on land. Must [be] very terribly important in the economy of the region.

Well, hoping you and Gwen are the same—in the matter of turning over sand I mean, I am, yours very sincerely,

1941–1942 / 145

To Heinz Berggruen
May 2, 1942

Dear Heinz:
Toni had your letter in her file for answer, but if I know the gal, it may be sometime before she gets around to it, it may be in fact never. And I may as well assume myself the pleasant burden of corresponding for the both of us.

No reply yet from Jn. I'll let you know as soon as, or if, we have any dope.

Since then we had a couple of letters from Norm Carlson, as T probably mentioned. And I sent him a great bunch more of the dope on the mandated islands.

We spent a night at the castle, and I was tempted to make off with, as a loan, their copy of *Darkness at Noon* [Koestler 1941], but will catch up on reading, if ever, first.

The Kafka thing was something. Yes I am very grateful to you as the instrument of our running on to him. Bold and subtle too if such a contradiction can be used. And it can. I used it! I will read everything of the sort I can find. I wonder why more of us don't get to know the man better.

I am so hipped on Goethe—to me the peak of an age—that I regard all German thought, however bifurcated previously, as coming into him as unity. And all subsequent truly German work as stemming from him, again however bifurcated. And so I tried to express the subsequent ideas, in my shorthand to John, as a "fundament-diversion series," G[oethe] representing, to me, the fundament, and the others a divergent series in about that order. Of course. *Faust* furthermore (altho I know that it is regarded, and correctly, as humanistic, and of course in that connection you are quite right, and I used the terms entirely incorrectly) departs from reason at the very start. "Hence I know that I cannot know anything." ... in an intellectual sense. The redemption in death of Gretchen at the end of the first part is equally an a-intellectual thing, and Mephistopheles, as trying to play the game by the rules of reason, would seem justified at saying "no fair." But most of all by the ending in which he is again cheated, intellectually, altho Faust in fact—but not in spirit, a far deeper and less reasoning thing—by his conforming to the terms, delivered himself over. Of all things, *Faust* is most, to me, a refutation of reason and possibilities. As *Flight to Arras* [Saint-Exupéry 1942] is. Well, no matter. It's just a difference of terms. Probably we are entirely agreed in concept. In any case Toni just called in that if I would whip the cream we'd eat, which with the end of the paper, means the end of this letter.

> Heinz Berggruen (1914–) was in the army, stationed at the presidio in Monterey. Ricketts befriended Berggruen, who became a frequent visitor to the lab. Berggruen later settled in Paris, where he opened an art gallery and met Pablo Picasso and Henri Matisse. He presently owns one of the world's largest art collections—including eighty-five Picassos—which is housed in Berlin. Of Berggruen, Toni remembers, "He always did love art" (Volcani Interview, December 8, 2000).
>
> In his April 23, 1942, letter to Ricketts and Toni, Berggruen wrote:

Under separate cover I am sending you two copies of the *Partisan Review*, which, I am sure, you will find an exceptionally intelligent publication. I would like to draw your attention particularly to the stories by Mary McCarthy and Franz Kafka. McCarthy is a young American who, I understand, has just [put] her first book out. When I read the story, I was struck with its sense of awareness. As to Kafka, I don't know whether this will be your introduction to his work or if you are already familiar with him. Very few have read him in German, and fewer in English (altho practically all his writings are now available in English). To me, Kafka is one of the greatest writers, and if this is your introduction to him, I almost envy you for the experience of discovering him for yourself.

1941–1942 / 147

To John Steinbeck
May 2, 1942

Have just finished careful reading of *Flight to Arras*. Which I suppose you read. If not by all means do so. For several reasons. Because first it's one of the most interesting things I've read in a long time. An important piece of literature too probably; the architecture—except towards the last where he falls down a little in structure and gets to preaching—I think VG, sort of an essay. Conceptually significant, particularly as regards the factors and maybe the immediate causes of the fall of France. Heretofore I had never been able to understand how any country could disintegrate so suddenly. There's probably more light on this problem in that little book than I would expect to get from a careful study of the military and tactical history of that part of the war. Whether or not his analysis of the factors is correct, or whether those factors themselves were only a function of senescence, still I think he put his finger on the immediate weighing down. Something like the weight I felt in the army in the matter of sugar being poured into 50 gal GI cans of chocolate and then not being stirred up. Probably I told you. Made a big impression on me. I figured that when I got to be KP, I'd taken onto myself that function, since no one else saw to it. Well of course it wasn't long until I was KP. And then I did nothing about it. Surprised me terribly even at the time. The weight of that inertia was so great that it got me who was ordinarily pretty meticulous and thoughtful and not at all likely to be swayed by a deteriorating mass activity. The guy seems to be alright. And a mystic gets away with a best seller! Should think it might be valuable for anyone who wanted contact with the Fr[ench] mind, or rather, with what's perhaps "best" in the modern Fr[ench] nation. His vision must be particularly clear for him to see "what it is in that darn village we must die to protect" because a French reconnaissance officer during that disintegrating time could cover himself with the usual glory only with the greatest difficulty. So he had to get an unusually deep and permanent glory, or none at all. Such a person knew that the lives being sacrificed were needless, France couldn't be saved, it was already slipping away.

To Theodore Seixas Solomons
May 11, 1942

Dear Mr. Solomons:
Our little Toni—who answers to the name of Eleanor among her own people, but who apparently becomes no more reliable as a correspondent even under the old name—swears solemnly that she attended to the LA red-tape situation. And she may even have *written* to you. But I doubt that. Anyway, that attends to the letters you wrote us, coincidentally, about the time we came thru there.

We picked up a copy of *Tomorrow* on the way back from LA. And now a couple of days back we succeeded in getting the June issue here. Not many places carry it. The article by (of all people) Dorothy Thompson I thought was a fine job. *Partisan Review* (now defunct I understand) is another interesting magazine with which we've just got acquainted. German boy in SF, interested in Kafka, did the introducing.

Dick Albee just spent a week here. George Albee's brother. Who did the article on Boodin. All of which stimulated us to get busy on Boodin again, and Toni just brought down from Jon's place the 3 volumes he had up there "A realistic universe" 1916 revised 1931; "God" and "Three Interpretations of the Universe" both Nov. 1934. Macmillan, NY.

I bogged down on the new book Jon and I were doing. It has no timeliness and during these days my interest tends to get tangled up into timely things.—The good ones of which have timeless qualities in them, too. Reading a lot about Germany. And curiously the best translation of *Faust* (by McIntire, New Directions) as just been issued in these German-important-times. Also the hodge podge which is *Mein Kampf;* a curious job, a mix-up of screw-ball-ness and idealistic depths if I ever saw one.

I will some day find that letter of yours which you say, and Toni says, remains unanswered. Hard for me to believe, because I did reply to it, but I guess only in my mind. And since it takes lots of time and effort to write the long letters that seem to be the only kind I *do* write, I probably let the intent answer for the deed. And a fine roll call!

Best regards to you both

> Theodore Seixas Solomons (1870–1947) was a founding member of the Sierra Club; he pioneered the Sierra Mountains in the late nineteenth century and mapped the route that became the John Muir Trail. Solomons was Toni's father, and after a visit to Monterey in 1944 he corresponded periodically with Ricketts.

⌒

To Heinz Berggruen
June 5, 1942

Heinz Berggruen
No. 4, 25th Ave.
San Francisco, Calif.

Dear Heinz:
I suppose we have you to thank for the *Colossus of Maroussi* [Miller 1941] which arrived yesterday. Toni figures you must have heard us talk about it. Funny thing, I bought two copies of it, but gave both away. And we did very much want one. I rate it and *Black Lamb and Gray Falcon* [West 1941] and *Sea of Cortez* as three of the fine travel books of these days. Now we have all three. Altho we had to co-author one to get it!

I suppose the draft board-coordinator mix-up is still unsettled else we would have heard. That it still is pending is better than that it should be settled wrongly, but I suppose you are anxious to get into the radio work and to put the uncertainty behind you. Somewhere recently I read that no healthy man under 45 can plan his life now with certainty. I am wondering even if that won't extend to the groups over 45.

When I was at UC campus last week I ran onto a friend I hadn't seen for several years. Prof. Siegfried von Ciriacy Wantrup, now at the Faculty Club. Something like 4 or 5 years ago he came to UC as professor of agricultural economics from Germany where he had been in that work at one of the well known universities, I forget which. Had written a text which I understand is still being used there. Agreed in the main with what the Nazis were doing at that time, especially for the workers, but disagreed with their attempts to dictate policies to intellectuals, and got out. I had been wondering about him, supposing that he may have had time to become a citizen. But when I saw him there quite by chance I found that he had still a year to go, that he is rated as an enemy alien, and he couldn't even come over to SF with me when I suggested it. Said he could go if he had some professorial reason, and that he could get by if he trumped one up, but he preferred not to do this. Said he expected to be drafted. I asked him why he didn't try for some work, in or out of the army, to which his abilities adapted him. Better at least than being a soldier. In a struggle that demands of us the best we have in the way of intelligence, it hurts me to see good minds lie misused. We have plenty of brawn—what we need now is plenty of intelligence to direct that brawn, in economic and philosophic strategy as well as in military strategy. And we have it. But we don't use it. We don't even largely understand our enemy's method of thinking. A serious error that could prove fatal. Ridiculous even in the larger picture, because in science of differential, even tho all the differentials may have equal significance, we must choose the one to which we are most nearly related. And get behind it.

Anyway, Siegfried approved, and I told him I would try to get him some information on how and where to apply. Can you take time to drop him a line as to your own procedure—who you got in touch with etc.? Or send it to me, and I'll forward it to him.

The situation is that in Germany he seems to have had something to do with the evaluating and possibly with the applying of agricultural rationing. Seems to be very familiar with the exact amount of animal fat allowed per individual during the various years, etc., and had a considerable share in this aspect of totalitarian planning.

Might be able to apply equal energy onto this side, which he believes in, having chosen it of his own free will.

You might even like to know him. I think he'd welcome any contact right now particularly, since he's confined to the Faculty Club after 8 PM and since, probably, his morale isn't as high as it has been.

Anyway, if you can squeeze out the time, send along to him or to me any apropos information you have.

We look to see you here soon. There again, if your hands aren't already too full. And I can believe they must be. If you should come here this weekend, we'll be on Portola Road in Carmel Woods (north of the town) at my sister's place (she's going to the city and we'll be there looking after my mother), Fred Strong, Carmel 1146. Hard place to find, but I can meet you somewhere.

So—

> Berggruen wrote to Ricketts later: "I am so much looking forward to seeing you. [. . .] I heard from the Coordinator in New York. They have a job for me, exactly the kind of work I want to do and should do. (German propaganda analysis). However, I'm having a terrible time with my draft board" (June 1942).

To Heinz Berggruen
July 11, 1942

Dear Heinz:
Toni has already written you and is right now I believe writing again, so I haven't much catching up to do.

We are both of us grateful for being put on the trail of Koestler. A new light to me, or a possibly correct light, on the Moscow trials. Which for years have puzzled me. Wherein the ends and the means get mutually sullied. Altho apparently, according to that reason, not necessarily superficially—spiritually so. If there is such an idea. I liked the final working out—the idea of relation. The holistic-ness. The fellow must know something about Jung's ideas. The "oceanic sense" which also St. Exupery knew but under another name—or maybe nameless—is undoubtedly the same thing that Jung has in mind when he discusses the 5th or emergent psychic function. To which thinking, feeling, intuition and sensation all contribute, and in which they all merge. A super-intuition. The junction of Swedenborg's divine love and divine wisdom. The result of what I call "breaking through," for which there are many vehicles. And which elevates anything else or everything else as valuable and dear only as it has contributed to the breaking thru—the tao. But which at the same time enriches everything potentially, since potentially anything may be a vehicle for breaking thru.

Well. A long way from Koestler. But he must be familiar with Jung at first hand or thru the seminars, because I think that idea hasn't otherwise been expressed in publication. Or maybe the most recent Jung book has it. I haven't seen it yet.

Dan Gibb's book "Search for Sanity" just arrived. A good man. We know him personally, would not have thot of getting the book otherwise. And Toni is reading it. Which she finds a little wordy. A pity, since it's about the only intellectual or rather logical-approach statement of the Jung ideas. Don't suppose I'll read it. I don't do so well with the ponderous things. Probably because I have a tendency myself towards ponderousness in writing.

And also Boodin's *Realistic Universe*. And the Spengler *Untergang* [*The Decline of the West*, 1932]. A better word than Decline. That Toni. Soon she will have read that through. Anyone who reads a thing thru has my respect. I tend to read it only enough to become familiar with the dialect. A bad habit and one I'm trying to curb. The hard way: by reading Hitler through.

Doubt if the publishers would be cooperative in the matter of sending out some "Sea of Cortez" copies. They seem almost solely interested in sales. Altho you can't always tell. I'd be willing to send out 2 copies if they'd send two also. And probably Jon would. Or more. Jn and the publishers can afford it I should think a lot easier than

I, who tend to be rather generous and good natured about such things. I think I will write to them. Altho no doubt soldiers would much prefer something else. But curiously the Mont Presidio library had 3 complete sets of *Remembrance of Things Past* [Proust 1931]. I suppose most libraries take gladly what is given to them.

I suppose you noted that Koestler has a new book. Based on his experiences in Spain waiting in prison for presumed execution. Good that he got off. The world had something worthwhile saved for it then.

Well by now you will know something of your destiny in the army. Altho it may take them longer to place you than the average.

Rainy up here. And rainy. And rainy. As Puget Sd. country usually is. But with one really hot spell that helped my flagging beer thirst. The collecting generally good. We spent a while in Seattle this week. The Univ. of Washington library fairly good. Quite good. Not as good as UC Bkly, but that's exceptional, the best west of Chicago. At Oceanography and Biology and in the library all, I was well received. They are building up a good school here. Washington is almost a socialist state. And Brit. Columbia is even more so. The profits of the B.C. government liquor stores (which are like banks!) go to support the schools. A fine system.

Long letter from Jn who speaks highly of the air corp. Apparently an intelligent and well run outfit. Army seems to be picking up all around. Stream lining.

Toni a leetle mite knocked out with a tooth extraction. And with much chasing around, long driving, and not much sleep. "In this our life" is fine movie.

Well, we'll see you,

> In the summer of 1942 Ricketts and Toni traveled to the Puget Sound region. They explored Washington state and British Columbia and collected during most of July and August. At the same time,

Steinbeck was writing *Bombs Away: The Story of a Bomber Team* for the air force.

To Barbara Stevenson and Ellwood Graham
Hoodsport, Washington
July 13, 1942

Dear Barb and Ellwood:
Toni says I owe you a letter. I say No. It can't be because: In the 1st place, I never write letters, therefore I owe them to no one. 2nd: I have received none from you, ever. 3rd, why argue with a woman. But I could tear this up I suppose.

I am in the process of having the legs walked offen me. Or as some would say, the pants. Not an easy thing to do. So that Toni wouldn't get too fat on this easy life, we make a practice, or *she* makes a practice of *me,* of taking a walk every evening. In the cool of the evening. After the hot of the day. It (actually) being raining all the time. And while we walk, we botanize. We have got so we recognize one tree. Because this is an achievement of no small significance, I ought to record the name. It's the Douglas fir. All over the hills. And we practically never miss recognizing it. Still, it's all fun. I feel a little bit bad about having lost my taste for beer. But you can't have everything. I can, but you can't. Everything.

Some of the collecting places have been shut off by the army, and I'll have to ferret out new ones. Only to have them shut off too, probably, in time. Hoodsport looks just about the same. A sweet, dead town. After five years there are almost no changes. Even the same clerks in the store. Except that there is a dearth of young men. With a result that the radiance is such that I have to fan myself just the least leetle bit when I walk by some of the gals that are left behind. Behind.

Some would say this letter has little seriousness. But not us. Not any of us. Wild flowers are all over the hills. That is wherever Douglas Firs aren't. Sweet peas, foxgloves, fireweed, Scotch thistle, always spell Scotch with a capital. In honor [of] their chief export. And blackberries. Literally all over everything. Which also means blackberry pies. A woman down the road makes the nicest cherry pies you ever saw. I'd marry her but I don't think she's the marrying kind. Lots of Olympia oysters still. Altho the migration of the Japs has taken away the chief raisers and openers. To me the best oyster in the world. What Jn calls "the much over-rated oyster" for purposes wot you can easily imagine. Altho the reports we got from Gwen in New Orleans would lead you to believe the beast isn't over-rated at all, or else Jon nearly foundered on them. At one time Gwen rumored that she was even growing pearls, in interesting places.

I don't know if we'll get into Canada. There are some difficulties to traveling here, altho the gas shortage seems to be much overrated. We got to Seattle once and expect to go again. The Univ. Washington library there is pretty fine. Next best to U Calif. in the whole region west of the Mississippi. And I will get in a little research there. So we hope you are having fun too. And on the whole I'm glad T. forced me to write. She said: or else—crossing her legs.

To Horace "Sparky" Enea
Hoodsport, Washington
August 5, 1942

Dear Sparky:
Your letter of July 13th from Phoenix Ariz. chased me around a little bit—you also having changed address in the meantime just to make it interesting—and finally found me up here.

We'll leave to go back to the Peninsula now in a few days. First I want to see my youngest daughter. Now a young lady, as nearly as I can figure from her almost constant references to boy friends. In Bremerton. Where there are certainly many of them.

I was in Seattle too a couple of times. Gosh you'd be surprised to see the changes in the waterfront. Seattle is all lit up at night. Shipbuilding yards I suppose. Looks like Pittsburgh, or Gary, Indiana. And you have to have a waterfront pass with your photo and fingerprints even to go out on some of the wharves. And you can't get on a fishing boat without it.

The other day Toni and I (Toni's learning to drive; going to be pretty good in time) picked up a soldier outside of Olympia. He was a cook too. Liked it pretty well and made quite a little extra money that way I gathered, corporal or something of the sort. But certainly didn't like the long hours. He was married and got all night leave twice a week I think he said. But had to be back so early in the morning, and lived so far away, I shouldn't think it would be much fun. Of course tho it really is fun. Who am I to say otherwise!

That Tiny! I bet they get him in the army. Or better still: the Navy. I think he'd be a pretty good guy on the working end of a machine gun, especially if he was a little bit mad at the Japs. Wonder what happened to his wife. Heard they split up.

Carol lives in Carmel, on Scenic Drive. In quite a nice house she just bought down on the waterfront. Don't know the address. But I think a letter addressed to Carmel would probably reach her. Mail there is distributed entirely in boxes, and I don't know what number hers is. Jon is still in New York. I have been writing him upstate. But a letter addressed to him care of McIntosh and Otis, 18 E. 41st St., New York City, will get to him eventually. He's been chasing around the country for Uncle Sam too.

Ed Jr was in Rio Vista (over toward Sacramento) for a while. In Cal Pack Cannery. But now he's in SF again. My mother writes that he comes down to the peninsula every week end. I guess he'll soon be in the army. In the meantime he's earning pretty fancy money for a kid doing nothing but office work.

Good for Rose and Tony. I think it'll be a fine thing. They'll be fine parents. I can see Tony being the heavy father and getting away with it, and having a darn competent son to follow him on the purse seiner.

Stay with it Sparky. Don't let the Phoenix gals run away too hard or fast. I mean: hold 'em down.

Ed

Horace "Sparky" Enea was one of the crew on the 1940 Sea of Cortez trip. The crew also included Tiny Colletto, engineer Tex Travis, and captain Tony Berry, who owned the *Western Flyer*.

On August 5, 1942, Berggruen wrote to Ricketts and Toni:

[Jay] is quite unhappy at the moment. The army said they would ship him East to a training camp, regardless of his Navy commission. So he doesn't know what's going to happen to him. [. . .] yes, and I am still around. Still a casual, unassigned, four weeks today. It does get monotonous although there are wonderful, unexpected rewards in those fatigue duties with which my superiors have decided to keep me busy. [. . .] This morning, for instance, together with a shiny negro giant from the state of Louisiana, I was digging a grave on the post's cemetery. [. . .] He was an expert grave-digger (having done some graves in Texas back in '34) and taught me the secrets of making the grave "comfortable" for the coffin. The sergeant who came to inspect our job, said it was the trimmest grave he had seen dug in a long time. We felt very good.

To Heinz Berggruen
Hoodsport, Washington
Aug. 5, 1942

Dear Heinz:
Back to the correspondence again. Toni, in the meantime (I hope) having attended to her share of things. Which if I know her, she did by going swimming and such like.

A very sad letter from Jay [McEvoy]. Right after T wrote him at Monterey. You probably have heard from him too. He says you're to stay permanently at the Presidio. In what group? Personnel? Probably a good stunt. Monterey is really a pretty fair place. Of course not so lively as SF. And it will get deader and deader as more people move into the more populous centers of war effort. Bremerton is immense. You can scarcely poke a car into the town. What would it be if it weren't for staggering the shifts!

We've had a fairly successful and a very good summer. I haven't put in the time I hoped to spend on scientific writing, and on ideas of internationalism (which had better be thought about right now, before they're on us; or else as slaves of the Nazis or Japs we won't have much need of. Which is coming about alright said he, unless the United Nations get a few notions of good old fashioned aggressiveness into their heads, and abandon this eternal losing defense stuff. 2nd front to relieve the Russians, says he, and some real action in the Aleutians and in New Guinea).

And in the meantime, to resume a parenthesis, the Japs are consolidating their gains in the Aleutians and at Buna. Whence they'll stage a real all out attack against Dutch Harbor and Port Moresby. But the news from China is generally good. The US airmen are apparently doing a real job against great odds. And the Nazi drive in Russia is apparently slowing down. I hope their defense in depth extends sufficiently beyond where the Germans have reached.

Even at my advanced and cautious age I have been thinking about getting in things. But I don't know where or how. I'd rather do something contributory outside, where I could in any case be more effective. *If* only there were any machinery to use such as I.

Had only one letter from Jon since we've been here. He wonders if he may not get in too. I think it will be unfortunate, and his considerable talents won't be so wisely utilized. But the whole thing is surely what you said: a great pattern of waste. With everyone running around frantically wanting to help. Fine ergs of energy at the bottom. And, on the whole, good top leadership. But what rotten, conservative, in-between leadership! *Time*'s report that the Navy insists on monel metal urinals, and aluminum lockers and staircases, and the army, first class tinplate for their beer cans. Just because they have the priorities and can't see the large picture for considering their own perfectionist needs! Well, so are we all, I suppose.

That was a good picture: pointed I mean—of the grave digging. The way people are. The way we are, at least. And the way our organization is. Still, it's the whole picture, the large statistics, that count.

A letter from Bob Price (or did you know him. Probably.). He calls his place in Hawaii "Insulting Manor."

Hope to get over to see my youngest this week. And look forward to it. Haven't seen her now for five years. A young lady now. Interested in boy friends more than anything else in the world. Which I think is a fine indication. And will plan to see Jay on the way back. Figure on starting about the 15th, but many things can happen now and then. We'll see you.

↩

To Joseph Campbell
Hoodsport, Washington
Until August 15, 1942

Dear Joe:
Odd summers I'm likely to be anywhere but PG. Spending this month and part of August in the Puget Sound country. Forgot entirely about your possible plans for coming thru California, but if I had recalled, I should have realized in any case that Pearl Harbor changed all that.

We have had some mighty pretty collecting up here. Gonionemus chiefly. You recall: the little pulsating jellyfish that we took in the eel grass of Jamestown Bay, near where the *Grampus* [Jack Calvin's boat] was anchored.

And some unpretty weather. Pretty if you like rain. And a little exploring around, despite the gas shortage which isn't very serious here. I had five or six projects in mind. A couple of scientific papers to work on—the last echoes of *Sea of Cortez* work. And some ideas on internationalism to work on. Maybe a little war work. But not many of them seem to be getting done. University of Washington is working up to a fine school. The library is v.g. Next best to U Calif. which is better than anything west of the Mississippi and good in any terms. At that tho, in the west we must depend on libraries that you people back there would find pretty inadequate.

Yes, I'm afraid this winter may be very destructive. Collectivism is upon us, and better that it should come from within, even tho forced as a response to a world that will fight us that way, than that we should have it crammed down our throats by an outside enemy. The eternal verities will be with us still but they may take a form so new that those accustomed to the old lights may not recognize them. Altho I must say that, comparing the present with the witch hunting I remember from World War I, this country has been doing pretty well in the matter of sanities and kindnesses. That last sentence deserves a paragraph mark; it really has no previous connection.

Nevertheless I'd think it would be good if you got your ideas out, and if possible published. All the objectiveness and depth that can be made available ought to be. At best, and this is nearly best for a war, all belligerencies drag along a lot of biases. Tend to declare things as black and white that otherwise would be worked out as special cases.

Reading Koestler, [and] Hitler carefully. Probably a little of Dan Gibb's "Search for Sanity," Farrar Reinhart and pretty fair but I fear terribly verbose. Toni (girl friend, on the scene since you were here) getting into Boodin, and I fear will read Spengler completely thru and carefully, which is more than I've ever done. The copy you sent me burned up when the lab burnt, but I got the one volume edition since, and peck at it now and then. I'd say that it's surely the depth behind the Hitler dialectic.

I wish I could sit in on your two courses next year. Someone ought to abstract the whole thing, get the feeling of it so it can be perpetuated other than esoterically in the minds of the pupils. Oh incidentally, the McIntire (New Directions) *Faust* seems to me to be a fine job. Only good translation I've found. I hope he goes ahead with the 2nd part.

Say hello to Jean. Come on out when you can. Winter holidays maybe, if still there is any travel left for non-essentials in the war effort. *Us.* Altho merely as a balance wheel, such as we may be making a real contribution.

Ed

Ricketts composed the following letter on behalf of James Fitzgerald, who hoped to be commissioned in the navy but was not.

To Naval Procurement Bureau
August 24, 1942

Naval Procurement Bureau,
San Francisco, Calif.

Gentlemen:
James Fitzgerald tells me that he is applying for a commission in the Navy, and suggests that I recommend him to you.

I am doing this very gladly, and out of a considerable acquaintance.

I have known Mr. Fitzgerald for 12 or 14 years. During most of this time he has been associated with a local group who have been very close friends. Comprising Hilary Belloc (now in the British Navy), John Steinbeck (now engaged in war work back east), Ritchie Lovejoy (also in government work), and others, with wives and friends, all of us have had similar interests. Some of the ideas expressed in the recently published "Sea of Cortez" stem from the numerous discussions among us during this time.

In all these years, I have noted Mr. Fitzgerald's progress as an artist and as an individual with a great deal of admiration. He did my portrait 8 or 9 years ago, and I am sure that what he put into it was an important factor in whatever integration I may have achieved.

In a group not necessarily characterized by temperance—most Bohemian groups aren't!—Jim has been outstandingly moderate, usually abstinent. He has gone on ahead doing his work in spite of social diversions, and in the course of the years his output has been truly enormous.

It is certain that he can be rated as an individual of considerable significance both in a charactral and an outward sense. You will find him entirely competent, and a dependable worker of some enthusiasm. So far as I can see, he should fill all the qualifications for a naval officer and a gentleman, as I understand the term—and this speaks well both for Mr. Fitzgerald and for the Navy.

Sincerely,
E. F. Ricketts, President,
Pacific Biological Laboratories

To Horace "Sparky" Enea
August 28, 1942

Dear Sparky:
Back to Monterey (and cloudy weather—as tho we didn't have enough of it all summer!) I find your letter of Aug. 13. And the canneries going like mad. And a violent dimout over the Peninsula, especially over Carmel. Also a little trouble with this typewriter. Looks like somebody dropped it while we were away. Some of Ed's rowdy, rich Pebble Beach high school friends have been using the lab while we were away, for a whorehouse if you ask me, and maybe some of them bumped into it a little bit. Along with locking the safe open, and breaking about 50 phonograph records, and leaving moldy pie and spaghetti lying around. Even the nightwatchmen had fancy stories about the goings on. "Young kids" said the sergeant of police of Cannery Row, "some of the girls seemed to be about 16 years old." Well I don't see anything wrong with 16 yr. old girls,

at that, that time won't cure. Time and a little attention of the right sort.

From somewhere or other, about 40 purse seiners have turned up in MB. Some I have never seen before. I supposed they were all in the navy. Looks like there'll be a mighty busy canning season. Seems that the government took a firm stand right at the start of the season in the matter of strikes, and everybody's knuckling down to work.

You can be glad you're in warm sunny Arizona. Looks like we're not going to get any good weather here at all. Altho some people are foolish enough to sit out on the beach even so. Or even to go in swimming.

Toni and I went over to the Dan James place couple of days back. Went out in his fabulous catamaran. It's sort of a double surfboard. You wear a bathing suit and a heavy old coat or sweat shirt. Because it's cold, your seat will get wet and you may get wet all over. Then you guide the thing (which weighs only about 75 pounds,) with a couple of paddles. Like a canoe except lighter and more seaworthy. You can turn it over. Of course you can fall off. I never saw such a rig for going out among the surf swept rocks around Pt. Lobos and into the caves. Be fun to go out at night equipped with miner's light fastened on your head. But I suppose that would upset the sentries too much. They might want to fire at you just a little bit.

Hear from Jn occasionally. He may be out here, or at least down in southern Calif. next month. I have seen nearly everyone so far since getting back except Crl. Don't know where she is right now. Have to run over there. Now with the tire problem what it is, we're careful about running. Good for Tony. And you being an uncle. A process that should be often repeated; sometime it'll be a boy. And what the hell if it isn't. Girls are alright. Or so I've been told. I even used to believe it once myself when I was a lot younger. Said he, watching a lovely blond wiggling herself up the street. Oh incidentally, Flora Wood's place has gone thru a lot of trouble but it seems to be surviving them. Now it advertised room and board. I suspect somehow that it puts out room and board with added attractions! Which the Army may not approve of officially, but which the army men individually seem to like pretty well.

More and more of the local people are getting into service. Now Jim Fitzgerald, the local artist, has applied for a commission in the Navy. I'm not even sure of my own status. I'm over 45, but only 3 months over, and they seem to be grabbing everybody in sight. So I may see you someday, in which case give me an extra helping of that good stew that I've been hearing about. Toni says hello.

Ed

On August 20, 1942, David M. Clay wrote to Ricketts from the editorial department at Harcourt, Brace and Company:

Viking Press has given us permission to use a selection from your and Mr. Steinbeck's *Sea of Cortez*. We are eager to have a paragraph or so of biography to accompany the article.
    Would you do us the favor of sending in the enclosed envelope, as soon as you conveniently can, fifty to one hundred words about yourself, including your birth date and chief exploits? We go to press with the book very soon.

To Harcourt, Brace and Co.
New York, New York
August 31, 1942

Harcourt, Brace and Co., Inc.
382 Madison Ave.,
New York City, N.Y.
Attention: David M. Clay, Editorial Dept.

Dear Mr. Clay:
I have just returned from another lengthy biological renaissance and reply to your letter of August 20th has been correspondingly

delayed. But I probably wouldn't have answered it promptly anyway, so everything's just as it should be.

If the chasing of marine animals for the last twenty years can be considered exploit, there it is. And the publication of "Between Pacific Tides"—which however mostly involved dealing with publishers, an art in which I'll never become proficient.

For the record: I was born May 14, 1897. In Chicago. No doubt a curious birthplace for a person who expects to be concerned with the marine animals of wilderness shores. But presumably I didn't expect anything of the sort then. Later on, at the age of 6, I was ruined for any ordinary activities when an uncle who should have known better gave me some natural history curios and an old zoology textbook. Here I saw for the first time those magic and incorrect words "coral insects." Beyond this I can record only the facts that I am fond of old music (liturgy and madrigals), free verse translations of Chinese poetry, and (of course) cool beer on a hot day (it's not even warm here in P.G.). These also are curious anomalies, by which I must suppose that I shall be marked as a curious animal myself. And you may regard all this as evidence of a life equally curiously mis-spent. In these opinions I will concur, but only with the hope that the mis-spending goes merrily on its way.

If this wording sounds undignified, change it around. It's at any rate true and I think truly representative. As much as anything else can be representative in this uncategorical world.

I hate things of this sort as I used to hate having to write in autograph books, and probably do the one now as badly as I used to do the other. Anyway, the facts are there, and that's what you want.
                                        Sincerely
                                        E. F. Ricketts

To Horace "Sparky" Enea
September 16, 1942

Dear Sparky:
You're certainly moving around. Well, it's that way in the army.

A lot of old CCC camps probably will be rehabilitated before the war is over. Because they're taking in a great bunch of men, and they'll have to have some place to house them all. When I first got in during the last war, we were put in tents (Camp Grant, Illinois, about 40 miles NW of Chicago), and again when we were sent to Texas, we lived in tents. And whenever it rained, the company street was just a sea of mud. To get out on the inspection line without dirtying newly shined shoes was almost impossible.

I notice that one of our friends up in Oregon was pretty much in that fix, during the early days of a new camp. They had no hot water at first, and they were on field rations. Had no sheets, pillow cases etc. Those things get ironed out, but it's pretty hard on the guys who do the ironing.

Looks like I'll be joining you guys within the next few months, navy or army, if the physical exams etc aren't too stringent. Fact of the matter is that, at the present rate, there won't be any men under 46 left in civilian life by next spring. Wot'll the gals do. They'll all become Wacs, Waves, etc, with commuting trains running from the men's to the women's camps. Which won't be so bad either.

Toni has learned to drive and is out with the car now. She wants to get driver's license before I go, if I go, so she can jump in the car and follow where I may be sent. If it isn't out of the country. Old guys like me I don't think they'll want for duty in combat zones, but you can never tell.

No direct word still from Jn. But Toby Street said he had a card from Hollywood saying Jn would advise his address there within a few days. I haven't seen him now for nearly a year. Crl stopped in here once with the news about Tiny.

Roofers are working overhead. Got so we didn't have any comfortable place to sleep whenever it rained. About to ruin books, typewriters, etc. Things have gone up frightfully. This roof costs more than twice what the last one did.

Almost as many purse seiners working here now as there were in the old days. Don't know where they all come from; thought most of the boats had enlisted for the duration. Suppose there is merely a concentration of all available boats for the sardine season here. Some of the canneries have been going pretty heavily, but fish deliveries have been a little sparse recently.

Still lots of going on at the lab, empty beer bottles and suchlike, but the gang is different. A couple of boys from the Presidio turn out to be pretty likable fellows. They're working them awfully hard up here, seven days a week, and long hours.

The war is changing everything. Now there are women cannery inspectors for the Fish and Game Commission instead of men. There is going to be greater equality between the sexes than ever before. Well it's OK with me so long as it doesn't give the women any ideas about being able to get along without men.

You probably heard the sad story of the Lone Star [brothel]. Everything was moved out, including of course all the beds—and boy were there a lot of them! The front of the building was torn off, and now the place is being used to store fish meal. What a fate! Far as I can see now, there are no sporting houses around town at all. Only a few street walkers left (probably, I haven't seen any, but I wouldn't be likely to) and of course a host of occasional gals. Who, I suppose, can put out a better dose of clap and more of it, than any well regulated house could hope to do. Oh I meant to ask you, do you ever go to USO huts, and what about them? We hear all sorts of conflicting reports.

Write again soon Sparky. I'm wondering where from this time!

Ed

In September 1942 Steinbeck and Gwyn moved to the San Fernando Valley in California.

∽

To Cornelia Ricketts
September 17, 1942

Dear Cornelia:
What news you have! And what a good responsible job for you kids to be doing! I hope your mother gets along fine, and know that she will. See that she gets my Aug. 30th letter as soon as you get her address, because I want her to know that I certainly think she's been doing a fine job.

Work has been slow here, California schools mostly haven't even opened yet, but I am planning things so that if I get in the Army, the lab won't be lost entirely. And have been doing some work in connection with writing. I sent in the article that I wrote up there, first improving it a lot, but haven't heard anything. Probably take quite a while to get it published, if ever.

I'm glad you're having a full period of chorus every day. That sort of thing can be a great joy to you, even after you get out of school.

Toni's daughter Kay hasn't come back yet, and it looks as tho she may have to go up there to get her. And to have her ex-husband cited for contempt of court. It'll have to be done up there because the divorce case was decided originally in California, and Kay is now out of the state. So Calif. has no jurisdiction. But we may wait to see if I'll be drafted, or if otherwise I can get in the armed forces.

Barbara and Ellwood [Graham] left to pick fruit. They got terribly broke. And just at the last minute, Toni took some of Ellwood's paintings over to Lewis Milestone's room in Hotel Del Monte (he has been here with a movie outfit and Toni has been working for him), and he bought a $100 painting and had Toni send 6 others down to Hollywood so his wife Kendall could see them. Probably they'll buy more.

The movie they're making is "Edge of Darkness." Toni has been helping with the script, and she says it'll be a fine job. Ed Jr was very thrilled because some of the people came around to the lab. The writer, Bob Rossin [Rossen?]. And also because we had cocktails with Ann Sheridan. And Toni had lunches with the other principals, Errol Flynn, Judith Anderson and some others that I've forgotten. They all seem to like her.

Jon got down to Hollywood earlier in the week but I haven't seen him. Suppose he'll come up here. Supposed to work on some sort of movie for the government.

Toni has been working very hard, what with the movie job etc, and probably won't have much chance to write, but I'll pinch hit for her. Most of all I want to see that she keeps working on some stories of her own which I think will be publishable if she ever finished them.

We know a couple of kids in the Presidio here and they come over to the lab often. Also Toni and I try to keep up correspondence with a few others. Soon everybody in the country either will be in the Army or Navy, or writing to some one who is.

Well, best love to you good people. Write when you can.
Your father
Ed

In September 1942 Steinbeck began work on the film version of *Bombs Away*, which was never completed.

To Bob Price
October 7, 1942

Dear Crpl Price esq.
Airmail letters is good. Even tho sometimes something gets in the way of their immediate answering. Sometimes. This time it was a little matter of an impending induction which struck. So I'm in, despite my 45 yrs 6 mos, despite my business property, expenses of

which will go on whether I'm here or not—and Uncle Sam doesn't seem to want to pay them—and such relational lives as people of our hoariness will have built up. But war is never a respecter of persons and who am I to think that it'll respect me.

So for me twice to the mark. And all I can say is: I hope this army is better than the last one. And it seems to be. At least the details are efficient, whatever the large picture may be (and that seems to be not so good; great coordinations were invented subsequent to the laying out of the plans of most military staff ideas).

Toni is working on MSS correction. And the title of the slightly screwball thing she's correcting should be, by popular vote: circumstances alter ego. But I'll bet it won't. It'll be greenfields and fertile valleys. Little trouble with this typewriter. Mine has gone to the shop. If there were too many typewriters around here none of them would be working or words to that effect.

But it would still be better if Ed Jr were going. He's past 19. And as the cynics say around here he won't have to move out of his swank job until after the election. Now he thinks only of three things: (1) of swing music (2) of his job as bkpr in Mr. California Packing Company's plant (3) that he'll soon be in the army. Oh yes and maybe of charming young ladies. Just a little. Not enough to hurt. Not anything like the way I thought of them at his age. Or at mine. Or even at the not so tender age of 7 when the lovely (Miss) Alice Cockrell, I even remember the name after these nearly 40 years—whatever it was that she did.

Koffka [Kafka] got really into the heart of the problem posed by your motors. Did you read any of his things in the *Partisan Review*. Now discontinued I believe. The little German—probably you know him too—Heinz Berggruen (now in the army, rest his soul) translated a couple of his things.

One thing I must do first of all: send you my poem. Beautiful, beautiful poetry:

On a separate sheet. I discovered I had one copy left, so I needn't type it out again.

Some people which is bums (you might say) right now *are* playing Scarlatti. And very good it sounds coming in from the other room into where I sit here, working. Before that some of Boyce, continuer of Purcell. And the concerto grosso of Corelli which is out in a very good new Victor issue.

I still envy you those tidepools. Well who knows but that I may join you. I am Ltd. Service, acct teeth, feet and stomach, but Hawaii or Australia might be duck soup for properly limited service.

Since this is airmail might as well turn over, said he. We saw Virginia recently; during the burning out episode. As she probably wrote you. A purification process. But don't try to tell her that when she's sorting over the contents of a half burned clothese closet. I always spell some words a little wrong. It's good for the language. Helps it to grow.

We were up there in connection with my final fling at getting commission in army, navy, anywhere. Army won't give commission from private life. And Navy will give nothing or Lt. Commander, which I obviously cannot qualify for. Can I take a 200 foot boat out of the harbor. No obviously not. But probably I could back it into a wharf if you are interested and don't want ever again to use the wharf. Except for firewood.

So the number of dishes I'll be washing in the next few weeks you just wouldn't believe. I still think I could serve the cause better by my knowledge of the oceanography of some of the Pacific islands, I did get to be finally quite a specialist in the Palao Ids. (headquarters of the Jap govnmnt in the South Seas) thru their own biological literature, fortunately published in English and available in the libraries of a few American biologists. The free masonry of biologists continually amazes me. The Japanese pure scientists worked out and recorded physical, geographical, oceanographical and biological data on their islands, published it, and sent separates around to equivalent scientists just as Fisher would send his starfish paper to a Japanese starfish specialist.

There was a new edition of *Between Pacific Tides* coming up. And it was a dandy. Jon wrote a forward. I had a fine map made, and a color photo that was something for a frontspiece. But now Stanford won't publish it until after the war. And Jon and I had already sidetracked a lot of work previously started. Toni and I are going down there (Jn's in Hollywood, but under such strict Uncle Sam orders that he can't get up here at least until after I'm "in"). Sunday nite for a few days. Haven't seen him for a year.

Well Scarlatti continues. This night doesn't get any earlier. I will be sleepy. Let's hear from you again. Same address, Toni will be here for a while. Until I know where I'm going; then, both of us hope, she can hop in the car and come there. If not too far. And if finances aren't too bad. She figures now on getting full time job until then. Scarlatti and Scardigli get mixed up. And really there's some reason for it. Said he, paying an old fashioned and pleasantly courtly compliment obliquely. he he

> Ricketts's reference to "the problem posed by your motors" in the sixth paragraph was written in response to Price's letter of August 27, 1942, in which he wrote the following:
>
> I am sending a short story I wrote about motors.
>   Once upon a time one little motor made love to another little motor. "Happy little motor, put, put, put." "Put, put, put," murmured the other motor's happy bearings. They sat thus in the big factory making love to one another until one motor was outmoded and was moved away. The other little motor was unhappy. It started going, "Pufttf, pufttf, pufttf," because it was unhappy. They gave it a thorough overhauling many times, but it still went, "pufttf, pufttf, pufttf." It too was moved away. No one knows whether the two motors ever met again.
>   You can see my dilemma. Do you agree with me?

Ricketts did, in fact, find himself drafted in mid September 1942. He closed Pacific Biological Laboratories for business, spending only weekends and leave time at home. He served at the presidio in Monterey as a medical technician.

∽

To Horace "Sparky" Enea
October 17, 1942

Dear Sparky:
I am up to joining you guys. Monday AM I will be yellow fever shotted and I.Q.'d to someone's heart's content, if not to mine.

Apparently I am one of the first guinea pigs of the new group of men—the old guys. The limited service men who will replace the lusty young guys who have dates with black gals in New Guinea.

Because I had lots of partial dependents, 3 children of my own, an ex-wife and a present wife, but not married that's the rub—some physical defects, and more than 45 years, I thought I'd be out for a while. But no sir, Uncle Sam wants me.

Will have to close up the lab, except as residence for Toni and Ed Jr. But the taxes and insurance and interest will go on just the same.

Toni and I saw Jon for a few days. We went to LA on the train. He doesn't know what he'll do next, or how, or whether in uniform or not.

Canneries have been having a terrible time. No fish. The boats are frozen here. But the help is drifting away to LA where there seem to be lots of fish and not enough labor. A funny time.

So I'll see you Sparky. Maybe in a couple of years. Or maybe I'll show up somewhere so you can hand me out some of that good spaghetti.

To Nan Ricketts
November 8, 1942

Dear Nan:
I have been in the army now for three weeks.

For a week I have been in the Medical Dept., working in the laboratory at the dispensary. Fairly interesting work, but terribly long hours, and 7 days a week. However, in the lab you're pretty much your own boss, and it's an honored position, so I guess I'm lucky. They needed men terribly badly; they haven't even bothered to give me much basic training. I have to be there before 7 AM, and sometimes I don't get thru until 9 PM, but there are slack times too, and when I get onto the work I think it will be easy.

I will make you sort of a proposition. By having $28 per month taken out of my pay I can have $50 per month sent to you by mister uncle sam. In other words, the extra $22 will be free. Then by having an extra $5 or $6 taken out, I can have $12 per month sent to Cornelia until she is 18. The government will contribute thus some $27 or $28 per month to the support of my legal wife and junior children.

It seems a shame to miss this opportunity, even tho it'll mean that I will get only about $17 per month salary out of my $50.00.

If you will set aside some of this money for a divorce after the war, and will send me an $18.75 bond every other month, I will have this money taken out of my check, and you will have $62 per month as long as I am in the service.

There's only one requirement. That I send our marriage certificate, or a certified copy of it, to someone or other in Washington (I'll find out who) within the next few months. If you haven't the certificate, and I don't suppose you have—probably I had it and it burned up in the fire—you might just let me know the exact date and I'll send to Chicago and get a certified copy. Think it was August 19th? Right. But what year?

Since starting this letter, I found your very lovely letter to me of Oct. 19th. I am so happy you had a good trip. Probably your arm is all well. I hope so. Have fun, Nan good friend, happy times. I am liking you so much. You and the girls are fine. And Ed is fine.

Ed and Nan Ricketts were never officially divorced.

Ed Jr. lived in the lab until he was drafted in March 1943. He remembers, "Knowing that Uncle Sam would soon be breathing down my neck, I left the Lab around Nov./Dec. 1942, ostensibly to visit my mother and sisters in Bremerton, Washington. It worked out beautifully; I was drafted out of Washington state and that sweet Army unknowingly sent me right back to California for my basic training" (Ricketts E-mail, July 13, 2000).

1943–1945

To Torsten Gislen
Pacific Biological Laboratories
Pacific Grove, California
April 7, 1943

Dear Gislen:
This will seem like a letter from a long-lost friend. I haven't written for some very long time. I guess since US went to war.

We had your Christmas letter of Oct. 26th, and were most appreciative. I am glad that it came thru, and hope this reply goes thru equally promptly. I feared that there would be no communication with Sweden whatsoever. Dr. Schmitt, even before Pearl Harbor, tried to get in touch with some anthozoa specialist there in Sweden, oh of course, it was Dr. Carlgren, in connection with some Panamic anemones I was most anxious to have identified, but reported he couldn't even contact him by air mail.

I had a copy of our most recent book "Sea of Cortez" sent to you along with a great many other specialists who received their copies. Fear it may not have gone thru, or that something may have happened to it en route. I do hope you got it; it may be some fun for you, and in any case the bibliography of the marine biology of the Panamic Faunal province will prove valuable. I put a great deal of work into it, as much almost, as I put into "Between Pacific Tides" but in much shorter time.

Now I have hoped to be able to make a survey of the marine invertebrates of the Gulf of Alaska, the Aleutian, the Bering Sea and the Kamchatka region, but of course that is now out of the question. Maybe after the war. And then I will be reasonably well equipped to start work on a manual of the marine invertebrates of the entire Pacific coast. Could be a very good job. If I live long enough. And if I can find funds to sustain myself while I'm doing the job—which will be financially *un*renumerative.

Ed Jr is in the army. Nancy Jane has been trying to join the WAACs, but she is still too young. My brother in law [Fred Strong] expects to be called. Don't know if you met him. Nice fellow, but very quiet, and not too well, fear he'll have a bad time in the army. Finally, I myself, was drafted last October. So we are quite an army family.

I am running the laboratory of the dispensary here at Monterey. Very interesting work. Urine analyses, urethral smears, venipunctures for Wassermanns, blood counts and differentials, etc. All involving lots of microscopic work which of course I like. The lab (PBL I mean) is going pretty well to pot, but there's little demand right now for specimens from schools, and maybe I can pick it up again after the war.

I have even been doing a little bibliographic research on Pac[ific] coast marine invertebrates, per previous schedule, in my spare time, but there's little of that, and I haven't been able to manage much really serious work. And the literature problem is more acute than ever before. I have been compiling a polychaet library that I hoped to make complete for Pacific coast descriptions, but it's pretty nearly hopeless to pick up the few Vidensk, Medd fra dansk naturhist Foran papers here, and one of them is pretty important, CCA Munro's 1928 account of the Mortensen, letter wouldn't go thru. I wish science were more international, in connection with things like that.

I finally married again, very kind and intelligent girl, considerably younger [Toni]. Now in third year and we get on well so far. I saw Nan last year up north and was very welcomed and welcoming. Even Cornelia, the youngest, is now a grown up lady. You remember "I'll tell my gweasman on you!"

Well, I will say goodbye to you and to Mrs. Gislen, and to the family I have never met, but which I feel warm towards nevertheless. I hope we may see you again sometime. Greetings and friendship!

To Ingersoll-Waterbury Co.
May 8, 1943

Dispensary SCU 1930
Presidio of Monterey, Calif.
The Ingersoll-Waterbury Co.
(Division of Waterbury Clock Co.)
Repair and Service Dept.,
Waterbury, Conn.
Attention: Mr. Ingersoll (or Mr. Waterbury)

Dear Mr. Ingersoll:
I am in a difficult situation and I think you should know about it. It's on account of one of your watches. The top sergeant is a good guy. A little bit tough maybe, but not nearly so bad as the guy at Headquarters co. And the Captain is fine. A good bunch of men I say. On the whole. But it naturally arouses their baser instincts when I get in a little bit late. Or a whole lot late. Or too early like the other morning when the boys said I disturbed their best sleep. I was just bored, it being so quiet and all when I got there an hour early everything was very dark and nothing to do.

Anyway my wife gave me this watch for Christmas. Because after I got in the army I needed a watch more than ever to keep time. In the army things are very punctual. I am a married man, Mr. Ingersoll and sleep home as you probably do with Mrs. I., but only 4 nights a week as they say in the army, not enough but what can you do? And probably you know that you can't buy alarm clocks for love or money. Or maybe your house is full of them and you wouldn't know. You being a manufacturer of clocks so to speak. And not very good ones if I do say so.

So after Christmas I thought "Now I can get to reveille and retreat on time and maybe not get into so much trouble." Well, the watch didn't keep such good time. It's your Warrior KI, and very G-I khaki-colored. So we took it to the dealer and fiddled with it, and still it didn't run so good. So my wife sent it to your SF branch, though how she found out the address is more than I can understand. She is certainly a livewire that woman. And they sent it right on back, and for a while I used it and it only gained or lost about 5 or 10 minutes per day but you can easy allow for that. Until it got so it would stop. And maybe during the night, with you waking up and looking at its luminous dial but then not knowing the real time any better than if you didn't look in the first place. So we sent it back again, and this time they sent it clear back east and it finally came back all nicely repaired or so we thought, silly fools that we were. Well hell what's the use. It wouldn't even start to go this time. I say if a watch won't run at all it's no good; it's just simply no god dam good at all Mr. Ingersoll, no matter if you did make it or if it's very pretty G-I khaki-colored and looks swank as hell on your wrist or on anybody's wrist I don't care who. And now I think my wife is pregnant though you can't blame the watch for that, directly, anyhow.

So you can see how it is. Now what I was thinking is this. If you can surely fix it so it'll run for say a year or even for a few months fine and good, I'd like that fine. But if you can't, let's not kid ourselves either of us. I hate to be sending it back and bothering you, and giving you quarters (25c) and paying postage and getting somebody to typewrite a label, and with your men probably busy too and wartime scarceness of labor and your men working on a hopeless thing.

Just don't send it back at all. Hell it's no good if it won't tell time. Let it be a present to you from a soldier far away in California on the shores of sunny Monterey Bay. And no hard feelings on either side Mr. Ingersoll those things just happen sometimes.

                                Your very respectfully
                                CPL E. F. Ricketts 39108601
                                Dispensary SCU 1930
                                Presidio of Monterey, Calif.

On June 29, 1943, J. E. McConkie, director of research at California Packing Corporation, wrote to Ricketts. "We are now seeking [a] suitable qualified person with scientific education to fill that position and contingent upon your being able to secure your discharge from the army to reenter a civilian occupation, can offer you this position. We hope to be able to fill this opening by July 15 or Aug 1st at the latest, and hence would appreciate your advising us as soon as possible of your status insofar as availability is concerned."

To J. E. McConkie
July 6, 1943

J. E. McConkie, Director of Research,
California Packing Co.,
101 California St.,
San Francisco, Calif.

Dear Mr. McConkie:
On the basis of your telegram, which arrived just two hours before the 4 PM deadline, I was able to initiate the army process calculated to result in my return to civilian life.

The first steps have been approved, and it looks as tho I shall be discharged in ample time to take up with Cal Pack on or before August 1st.

This in reply to your letter of June 29th—the second letter—which finally came through on July 2nd.

        Sincerely,
        E. F. Ricketts

After his discharge from the army, Ricketts began work at California Packing Corporation as a chemist.

To Edward F. Ricketts Jr.
August 8, 1943

Dear Ed:
We got your APO number card last week, and today, very welcomely, your wire. I called Frances immediately who may have worried about your trip, and now needn't.

I suppose you were glad to go, and in a way you are lucky to be in a foreign country, and that much nearer the heart of a very important thing. I even got to thinking favorably about places like New Guinea; I was thinking how much fun I'd have running a malaria lab during working hours, and looking at the marine invertebrates during spare time—if we were on the coast, or going there for days off. But I was pretty old for that, and it would have upset my family life, my scientific life and my financial life—such as it is—too much.

But I do envy you your chance to look over the animals. See and record everything you can; you'll have to be seeing for me too, because probably now I'll never get anywhere off the Pacific coast of NA. At one time I thot I might have a look at Hawaii, or the Grt. Barrier Reef of Australia, but now I know my life will be already too short for me to finish up the Pac. Coast of NA invertebrates. You can depend on local museums, if there are any, and in Australia there are a number of popular books on the Grt. Barrier reef animals.

Now out of print, but very interesting and readable, and available at any large library is CM Yonge's *A Year on the Grt Barrier Reef,* Putnam, NY 1930. There are a couple of other good general works, Russel and Yonge *The Seas,* etc, cited in the Ricketts and Calvin bibliography. Also publ. in Australia, by Angus and Robertson Ltd., Sydney, is the popular *Wonders of the Barrier Reef,* by TC Roughley, 1937. I should have this. If you are in that country, have a chance to get it, and will have a chance to bring anything back with you, by all means get this for me.

Now I am out of the army and working in the chem. lab at Cal Pac. Jn is in London, writing for the NY *Herald Tribune,* and his articles are syndicated; I read a few. In this region they're running, of all places, in the SF *Examiner.* Marge and Frank Lloyd have come back to Carmel. He is working for an electrician, she's going to have another baby; her fourth. I had a letter from Cornelia.

The work at Cal Pack is going to be OK if I can speed up my techniques of using the chainomatic balances, etc, fast enough to handle the full load. So far the cannery has been barely operating. I have learned the processes for determination of protein in meal, of free fatty acids in oil, and of moisture in oil, but do them very slowly. Have yet to learn the stickwater processes. Norm is fast and accurate; quite a competent fellow I should suppose. Cal Pack say they hope to be able to use me for my biological knowledge also eventually, but I can't see now how my curiously specialized training will be of value there. They're paying me $225. I requested transference to ERC status in the army, to work in defense industry, only 2 hours before the final deadline. Glad I did; things at the post got worse and worse in shifting around of personnel and increase of work. Toni is working in transportation, but has tried and continues to want part time work. Suppose the only way she'll get it is by taking on sets of books to keep etc. Kay is up in Portland for a few weeks vacation and to see her father. Bert Bayles comes over still often. I wish you could have met him. Lou Watter's piano player. Has a fine bunch of records, I guess fully 50 of the old Jelly Roll Morton piano and orchestra records among others. The Hot Peppers etc. I've got to like some of them very well. Winin' Boy Blues, etc.

Well, write as you get a chance. Frances probably will be writing too. We are going over there now with this letter so Fr can weigh it up for airmail or otherwise determine how it will reach you fastest.

Love

Edward F. Ricketts Jr. was nineteen at this time; his twentieth birthday was August 23, 1943.

Steinbeck went to England, North Africa, and off the coast of Italy as a war correspondent for the *New York Herald Tribune* from June through October 1943. His dispatches were widely syndicated; a collection of them was published in September 1958 as *Once There Was a War*.

∽

To Don Emblen
November 3, 1943

Dear Don:
Glad to hear. Pleasant for me if I devote what time I can to the discussion you suggest, but doubt if I can scrape up very much free time. Anyway this letter goes.

Have been spending all my work time and some of my spare time in chemical work and study, but now the fishermen are struck, work is light, and I can revere the procedure. Toni has a new and better job as credit mgr at the *Herald*. Gives her fewer hours work, and arranged pretty much as she wants them. Janko [Yanko] is back from the boat, where they had bad luck; now down in LA for his exhibition. Your speaking of Henry Miller reminds me to say that Janko met him down there, and Miller speaks of coming on here again. I think he is a good man. Charles Fort makes my tired ache, altho I realize I am one of a minority. Many people whose minds I respect admire him; Janko; at one time John; Toni. Most of the writers whose work appears not to be circumscribed by form are those who have got to use it as familiarly as a person uses his senses. And I suppose most of them went thru a conscious struggle on that score before finally getting into their own peculiarly characteristic and individual form, or becoming free of it altogether. I think that your poetry shows as a rule an automatic and probably unconscious adherence to good "form" or taste, something of the sort; in such ways you're a "natural," you may hit the thing at first attempt without having to go thru that struggle consciously. A radio technician whom I used to know found that he got finally to the point where it wasn't necessary for him to envision, then draw out, and then build a circuit in order to see if it was going to work; he got so he could accept them or reject them in his brain before even writing them out: He blue printed and worked out only the successful ones. I suppose some workers would go so far as to do all this unconsciously, only the acceptable product would emerge to be worked on. John S apparently works that way. He may not even know what his characters are going to do, bides his time with considerable interest and in the meantime tries only to keep clear and keen for receptivity. Probably what significance there is in this emerges actually not because of form, but in spite of form or no-form. But there's certainly plenty of evidence of significance emerging from a vehicle that is strictly disciplined in form. Keat's "Ode to a Nightingale" stands up well even in these times of freedom. And of course Flemish and Byzantine painting. Best of all Chinese poetry. The Ebritt [*Encyclopedia Britannica*], Eleventh to 13th Ed. has some noteworthy remarks in that connection in Giles' article on Chinese literature. After stating all the strict disciplines that constrain the construction of the 5 and the 7 syllable line, he notes that the product of this strict form often sounds so free that it seems almost as tho no trammels whatsoever restricted the poet in his choice. The same notion is reflected in old saws to the effect that no one can command so well as the person who can obey. And the same thing a person can observe in everyday life. No one is so free in table manners in an old fashioned group [for instance], as the person who knows perfectly the manners of his time, and how and where to transcend them gracefully. He actually is the only one who is free of the table manners, not the gauche youngster who violates them from ignorance. I was realizing that the other day in the restaurant. Juke box was playing some popular record. Trumpeter soloist wasn't keeping with the time of the background. He was instead creating around that background. But no one could do this except one very familiar with that background, who could play it, even compose it. He had to know just exactly what that background was in order freely to play around it so rightly.

There are a few lines in Miller's *World of Sex* also pretty much to this point, but I haven't the MS here and there's no sense to my trying to remember.

Well anyway, for me it all sums up: The only person who is free of form is the person who's been through it, knows it, and has come out on the other side. A mystic might say that such a one had super-form, or had emerged from it, as Walt Whitman, or Lao Tsu. Actually I think Whitman has a very close form; fact that in different editions he changes around single words and then puts them back as they were shows that his ear is feeling around cautiously all the time. But he achieves it mostly unconsciously, as in many of your things you do: a feeling for form rather than a driving after it.

This letter is a good example. If John, knowing what I do at this point, had written it, there would be no tautology, everything would be stated progressively, and there'd be no confusion. I could do that too but it would take several rewritings and maybe half a day's work as against Jon's half hour. Well, it's all good clean fun I suppose, and poor inefficient guys like me, who keep their muscles whetted by that very clumsiness, probably have just as much fun as the efficient people. And they certainly have less time to waste.

I hear from Ed Jr. occasionally, happy in Australia. You are lucky to be still on this coast. Glad I had this chance to write. See you

> Jean ("Yanko") Varda was a Greek painter who had been in the United States since 1939, living in Big Sur with his wife, Virginia. He and Henry Miller maintained a close friendship during Miller's stay in California.

⌒

To Xenia Cage
December 11, 1943

Dear Xen:
One thing you can say about writing to a person who never answers letters: you get a prompt and efficient answer. Tal for instance, definitely is not the type who answers letters, and Ed Ricketts isn't, also Xenia isn't; so I can't guess what all this proves.

Well, first for the news: R and T and John [Lovejoy] are fine; we see them every Sunday; Toni does her weekly washing in Tal's machine, we sit around and drink beer in the sunshine—when there is any sun—and have dinner there. They have a lovely little place out in the woods near the 17 Mile Drive entrance. Tal is going to have another baby; that's her secret but since she won't tell no one, someone has to say so; hell you have only to look at her to know it. Also speaking of babies, Marge Lloyd is about due. We were over there last night; she looks mighty sparkling too. The Lloyds and the Whittaker[s] and just about everyone else here has been working in the shipyards for fabulous wages, but most of them are drifting back, having concluded that the game isn't worth the wages. Jim and Peggy Fitzgerald sold their house and went back to Monhegan Maine, but since then I have got [a] request for recommendation for Jim from some defense plant back east, believe in Waltham, Mass.

Ritch is not in the army and hasn't been. He worked for the FSA long time, ran a camp down in the desert (Blythe), now working for the *Penin. Herald* as circulation Mgr.

I however have been in the army. For about a year. Now out again at my own request. In some ways I led the life of Riley. I had very long hours and sometimes quite hard work, lab technician in medical laboratory here at the Presidio of Monterey, urine analysis, blood counts, malaria work etc, all very interesting and right down my alley. Also I lived home all the time. More or less subrosa. AWOL 3 nites a week and out on legal pass the other 4. But I got a lot of fine understanding out of it again, my second time in the army, it reemphasized for me the impossible and the both large and intimate foolishness of any American collectivized life. Differed however from the preceding war which was both inefficient and dishonest in that I saw this time nothing but the strictest honesty. Of course a lot of little inconsequential graft, but this time the whole structure wasn't ridden with official graft.

Toni for a while worked at Ft. Ord in personnel, then took over Ed Jr's very well paying job at Cal Pack when he left to go in army, then went into Transportation at the Presidio, only a little while before I got out, now Credit Manager at *Peninsula Herald*.

I was able to be transferred to inactive status in the army only by taking up some defense work, so I have become a chemist. I am working during the day at the very beautifully equipped Cal Pack laboratory; tend to the lab work here, which is of course greatly curtailed, during the evening. And attend to such drinking as has to be done. I can thank the healthy army life for putting me back to the point where I can drink every night.

Ed Jr is in the Chief Signal Officer's office at SWPA headquarters in Australia. Last we heard he also was leading a very fortunate life—not yet fighting said his father earnestly knocking on wood—working on chart, I suppose of the Jap Mandated Ids. (incidentally I did a lot of this before being inducted, got to be quite a specialist on the Palau Islands, where the Jap Govt maintains a marine biological station, turned over a very good map of coral reefs etc to army and navy intelligence).

Both the girls are here now. Cornelia came down about two weeks ago from Bremerton, where Nan is selling her houses so as to get something in south Calif. where the climate is more suitable for her sinus. And last week Nancy Jane, who quit Bremerton Navy Yard and is trying again for the Canadian WAACs. They are fine people, we have lots of fun. Both the gals have really fine voices, good ears, with Toni that they go around singing madrigals and carols.

Toby divorced Frances (Street), married a young, Junoesque and slightly caustic blonde; they get along not too well, but there's probably more spark than there ever was in his former marriage. Barbara (Stevenson) believe they are married now, and Ellwood Graham (non-representational painter who did that fine portrait of Jn some years back before he went in more strongly for abstracts) are coming back here soon from Taos.

So I guess that's the news. Highlights of it anyway.

I'm glad to hear about Joe [Campbell]; glad that his fine mind is producing. I got an announcement of his Indian Painting book and sent for it immediately for Toni. Building up a good library again after the fire loss. Got the very good but slightly stuffy *Arts and the Art of Criticism* recently of Prof. Green (Princeton Univ.). And lots of biological literature. I am going ahead doggedly working up data on invertebrates of Pacific North America. Now on the Gulf of Alaska, Aleutian, Bering Sea, Kamchatka area where I hope to be able to explore within the next few years. If and when, I shall have covered the Pacific Coast pretty thoroly, and will be ready to take up in earnest the project of my definitive manual. Only problem is when to do it, wot with the exigencies of earning a living—or what, with us, passes for a living.

You know Xenia, once you told me (and gave me a much needed lift) that I was good people, and no danger of my losing touch. Now I tell you that even more truly. You can't lose touch. Whether I write or not, whether Tal writes, or whether even we see you, in one way isn't important, you are there just the same. And I am—with love.

To Nan Ricketts
January 4, 1944

Dear Nan:
Herewith deed, all notaried. Also I received a recommendation blank from Sears Roebuck, Glendale. They had you listed as PBL employee, wanted to know how much per week salary etc. I filled it in general terms, said your position had been general, associate; cited no salary, said you had been known here for many years, collecting specimens, etc, and that you were considered financially responsible.

Had a lovely Christmas with the girls, as they no doubt told you. Also had recently a nice letter from Ed, still in Australia.

Jn is back in this country; I am writing him tonight.

Good luck to you Nan; have fun; I hope this year and this climate sees your sinus better. Love to you all

Ed

> After returning from overseas in October 1943, Steinbeck began working on *Cannery Row*. In early January 1944, he saw the film *Lifeboat,* for which he had written the original screenplay. He was very unhappy with Alfred Hitchcock's alterations to the story, particularly in the portrayal of the "stock comedy Negro," which Steinbeck had originally conceived of as "a Negro of dignity, purpose, and personality" (Steinbeck and Wallsten 266). On January 11, 1944, the Steinbecks traveled to Mexico, where Steinbeck planned to work on his story idea for *The Pearl* and continue writing *Cannery Row;* yet Steinbeck spent much of the trip resting, and he and Gwyn returned to New York in March of that year (Benson 543–44).

To Nancy Jane Ricketts
January 5, 1944

Cal Pac 1-5-44

Dear Nancy Jane:
Glad to hear. Bee I bet is full of movies by now; and so are you. And full of rain if your neck of the woods is anything like this.

I will enclose a ck. with this letter to cover money you advanced for Cornelia's fare down there. And I will send on your navy check as soon as it arrives. You needn't wait for it if lack of finances is what is holding you up. I can probably dig up what you need and will be glad to do so. Let me know at once.

I have your mother's quitclaim deed all executed and enclosed. It will go out in the same mail with this letter or before. If it fails to arrive, let me know, because those things are important and should be attended to at once.

Nice letter from Jn. And very funny; said among other things that he and Gwen got nicely "empunched" couple of times over the holidays and I could just see that bowl of punch sitting there on the table and Jn practically climbing in it, only slightly restrained by Gwen who had something of the same idea herself.

Haven't seen Mother since you left. Toni went up to Burlingame to pick up Kay on Sunday, which is my usual day for visiting Crml. She left the car down at the depot; it was raining; R and T came over, we got to working on a little beer and result was I never did get the car and it was still there 7:30 PM when Toni and Kay arrived at depot expecting to be met. By a lot of rain and a disgruntled parked auto. Well, it started anyway.

One way of getting letters is to write three for every one you expect. Except for very well known people like Jn who never write any and are nevertheless deluged with them.

Can't think. Well here's one for you. Two men had a dog. Name was white. What color was methusela's grandmother. And how old was the dog. I had a little trouble making that one up so don't ask me for the answer.

Hello to Cornelia and your mother.

~

To Edward F. Ricketts Jr.
January 5, 1944

Dear Ed:
Your letter of Dec. 8th was received with a very warm welcome. In the meantime Frances had heard from your pal's wife that you were incommunicado, taking shots, and probably destined for the close-

up fighting front. I suppose that's part of the pattern, and something that most of the younger generation wouldn't willingly miss, but this older generation don't like the idea so well.

Soon after your letter of Oct. 12th came in, I replied to it. But for some reason failed to send the letter. Now I find it with a bunch of other stuff in a file that should have been attended to and send it on now.

Norm is at the moment in NY; Lyle Fosmark and I are holding it down and for a while it was a job. As much work, I gather, as this lab has ever had. Both the liver oil and the sardine end of it have been heavy, but now taking it easy again—or I wouldn't be writing this letter.

Jn as I may have mentioned is back in this country. Going to New Orleans and then to Mexico. He plans not to come back here on the coast until after the war, but I imagine we'll see him sooner even than that. Still thinking about and half planning some trips up north, clear over to Kamchatka, but don't know if I'll ever get there.

We had one of the nicest Christmases in recent years. Both girls were here. We all had fun. There was a succession of parties. I got Toni a particularly lovely present, only decent one I've ever given her. A Chinese (late Ming, around 1600–1700 as I recall) green agate snuff bottle that I had modified for use as a perfume bottle. And a portfolio of fine reprod[uction] of Navajo ceremonial paintings, with a commentary by Joe Campbell. Possibly you remember him. The Col. U guy that went up to Alaska with us.

Well well a little work to do, so I'll get at it. Have fun. And by all means have a look at Grt. Barrier Reef if you get a chance.

Love
Dad

To Remo Scardigli
January 14, 1944

Cal Pack
1-14-44

Dear Remo:
Now that Toni is facing the fact that she's simply going to sit around and look beautiful—which she does reasonably well—and realizes she's just simply not going to write letters, it seems that I may as well take up that burden which is no burden.

We had a very welcome letter from you and then a Christmas note. Implication seemed to be that you were in England, great secret, and occasionally at least in London. Now in another few days I expect to see notes from some foreign correspondent to the effect that you were seen in Ipswich-on-the-Stratford ascrewing of some blonde. If I know you. And if I know her.

I suppose Virginia writes you the news. That I am out of the army acct physical decrepitude, extreme age, bad moral effect on the younger men (they say I drink, and not alone). And working in defense industry—laboratory technician in Cal Pack Chemical lab only a few blocks from the house. Keeping up a little with my zoological work, but very little. Work here usually keeps us pretty busy, matter of fact I have to plan things very carefully and work efficiently and at considerable speed to get done in 8 1/2 or 9 hours ordinarily. But right now it's quiet, and so Ed Jr (leading the life of Riley—knock wood—in Australia) and the girls, and Remo gets wrote to.

We had the usual session of parties over the holidays, and some of our friends were so unkind as to say I overate. New Year's eve we were at Fr Elaines [Francis and Elaine Whittaker] as in the old days. Toni wrote up the party for social notes in the *Herald* (where she's working now as credit manager—can you imagine that! A gal who's never been known to pay up her own bills or mine either!). And at first the social editor was greatly relieved thinking "I'll just run this as is, save me writing it up." But when she saw the slightly personal nature that had to pop out acct Toni's slightly personal nature, they say she thot different. Since our little T-gal didn't gloss over the fact that too much drinking got done, and that one J Hopper's daughter bless her heart, having trouble with fly of her waist— which meant that breasts would pop out if they'd been big enough to pop—dispensed with clothes completely between there and the next place, so they say.

Well Remo some work to do, so I'll cut this off altho otherwise probably I'd have written more and more.

⁓

To James Fitzgerald
February 24, 1944

Dear Jim
Probably you thought that you wouldn't write, and that I never answer letters. I don't know how that reputation got started. Probably on Sunday, or under your roof which never leaks.

Speaking of roofs, or snow or something of the sort (you know it's been snowing so they say in LA, and El Toro here is covered); someone put his or her foot thru our roof on the shed in back. And it's raining. The other day out in back when I was trying to pack a few sharks I certainly knew moisture.

But most of the time as Toni probably told you, I'm a chemist. That it should ever happen to me who am not famous for liking mathematics. To have to use a computing machine at my age! Or did she tell you. If she did write, was it decent, and did she keep a carbon copy. Even she can't remember what she said. Swears now she's really going to keep carbons. Wonders if she told you about being arrested for s . . r . . ing Santa Claus on Alvarado St. in the rain. Or no, it was about her having that fellow arrested. Funny thing. I had gone to work, about 8 AM. Some fellow parked his car downstairs and started looking into the basement, then walked down there, cautiously. Then came upstairs, tried the door. After our adventure with that malicious and gigantic prowler 2 AM before I went in the army, she was concerned. So, without giving him the benefit of a doubt, she called the police. I hope putting some clothes [on] first, altho I don't know what difference it makes. Police probably like no clothes as well as anyone else. Anyway it worked out finally that the guy had just started work for [Max] Schaeffer, who refines shark liver oil in back of our place. And this fellow had never been here before, and Schaeffer was late that morning, and he was just wondering what to do and where to go, I hope. Toni was properly chagrined.

Well the straight news is that. Jn and Gwen were in New Orleans after NY, and now probably in Mexico. Covarrubias wrote Jn that a pre-Maya city had been discovered in the state of Vera Cruz, something like that, and he's probably there. Crl is around the Penin[sula] now that her Lt. Howard has gone for month's course back east. That marriage apparently took. The Lloyds are back. Marge had just had another baby. Frank told us only yesterday. His back bursting with pride. His 6th or is it his 8th. Girl. 8 1/2 pounds. Named Margery Morgan. Morgan le Fay say I—if she's anything like that beautiful Jennifer's growing up to be. Also the Whittakers are back. Fred Strong not yet called in army but expecting it any day. Ritchie Lovejoy has been classed 1A; he hoped to get delayed long enough to welcome their 2nd child due in May. Did you see John Cage and Xenia in NY. Had a letter from her awhile back. Ed Jr my oldest is T/5 in Signal outfit, in Australia. Something about map or chart making. Having [a] fine time, expects to be there for [the] duration and doesn't expect to do any actual combat work. Hope he's right. Nancy Jane tried again for the Canadian Waacs, was here, went all the way up to Victoria, then Vancouver, [where they] told her (after again soliciting her by letter) that Washington didn't like, new ruling [sic]. She's still a bit too young for US Waacs, so suppose she'll get [a] job here and settle down for while. Cornelia, the youngest who likes and holds her liquor, was here for awhile also. And the welkin was certainly ringing. Those gals can sing. Toby and Peggy stop in from time to time. Maybe that marriage will take; certainly never supposed it would at the start.

Card from the Cohees. See Dick and Jan Albee occasionally. He was classified finally as 4F. Andre Moreau has been going around with Margaret Lial; wonderful hand holding into-each-other's-eyes-looking, acc[ording] to Toni who has seen them in action.

Just had a letter from our Capt. (Med Corps). Makes me almost a little sorry I didn't stay with it and go overseas. He's in the Fiji Ids. Malaria control officer. Writes that if I were there, there'd be plenty of spare time for playing with marine animals—they are near the beach, and lots of fine malaria bugs for laboratory hours. Most of the rest of our gang got discharged on medical certificates. Some underweight, one for high blood pressure. One guy was too fat. I only, the oldest one of the lot, seemed to be in the pink of condition said he knocking on wood.

So anyway. I am working 8 hrs day more or less at chemistry, doing occasional work on future books, plans for trips, keeping up PBL after a fashion, working on Mandated Islands data, drinking a little, going out not so much as usual, and generally doing everything as usual except sleeping. Seems not to be much time for that. Now if you'll only teach me to paint, that'll look after *that* six hours, and my 24 will be full. Oh yes about Hilary. He has been here couple of times. Toni is fascinated at being around. Says each of us talk so fast that, not only can no one else get a word in edgewise, but that we anticipate each other's thoughts and finish each other's sentences, so that to listen to such monkey business for several hours makes nervous exhaustion. Anyway we settled the war a couple of times. Which needed doing. If they call you they'll get me back. I'm still in technically. Have never been discharged. And they can call me at the drop of a hat. Whenever they need a lab technician no doubt. Haven't seen Evelyn. Had a letter or two. Toni calls her up whenever she goes to the city, but E never home. Can you still get gas. Or do you still have the car. A car. We do alright since I walk to work. And Toni working on the Herald, has B book and gets what she needs. Well there it is Jim, there's all the news. And I'll bet you'll be as surprised to get this letter as I was to get yours. As we were to hear from Peg. As we were to get the announcement of your show. How did it go? Well received? And did it sell anything? I supposed each is important as the other. Did you get a job too? I got a reference about you from some outfit in Massachusetts. I was very Rotarian and coy and business like in replying. Assured the people that to my best knowledge—and I should know!—you were *sober,* industrious, gentile, conservative, easy to get along with and particularly liked fireman. Oh yes and one other thing what was it. That you were sober, that was it. I said you were sober Jim. So help me. Was it for a job as a fireman? I told em you were hell at putting out fires, especially after you'd been drinking beer. Toni says "beer, sssss" what does Pegs say about firemen, beer, Massachusetts and the gulf or Maine. Hell I probably can't even spell Mass. Lucky I don't live there

Ed

The following letter was written after Xenia left her husband, composer John Cage—known to be homosexual—due to infidelity.

To Xenia Cage
March 14, 1944

Dear Xen:
However [an] unexpected thing to you, it was much more so to me. I suppose those things have to be, but they're almost impossible to take at the time. Like a fever dream I used to have as a child.

At this point a 24 hr. interruption. So. I folded this up and stuffed it in my pocket and took it along to work. But had no time there either. And now it's back where it came from. Letter from Gwen in the meantime, Cuernavaca Mexico, very happily pregnant. Which reminds me of Tal. I took your package over to her and they were all as delighted as children. That perfume incidentally is something. I'd say it fulfills very adequately the purpose around which it probably was built.

I don't know their mail address altho I could take you right to the place. Tal ought to be spanked for not keeping in touch, particularly for not acknowledging the Christmas things that gave them all so much pleasure. But, actually, she has her hands pretty full. They live quite a ways out of town, Ritch works long hours, Tal does all her own washing and I suppose she gets so far behind with housework that it becomes a drag and her morale goes down. I will tell her about all this, take your letter out there and maybe that'll steam her up to write. As an expression of affection and support, letter writing is quite inadequate, but it's the only tie between people who are separated, it's better than nothing, and it can be [a] good thing. Still, look at what the Chinese did. Some of those poems undoubtedly were written to send as letters. "If you are coming through the narrows of the river Kiang." Pound translation is still best for that one, though erroneous. I have three or four others—other trans. of that I mean. Incidentally I suppose you know "Black Marigolds"; I think probably the best love poem "One night of disarray in her green hems . . . outweighs the order of earth."

I should think you might like to come out here. But if you have a sense of needing to determine something first of course that's the thing to do. Seems as tho a thing of this sort has to happen to everyone, to every good person anyhow. And you can't tell how you're going to come out of it. [A] person could make a mistake that would repercuss for the rest of his life; or until he made a life that had a different direction entirely. Impossible to pick up any decent advice. I suppose the only thing to do is to be as open as possible to any illumination that might come along. In things that concern a person alone—in his work anyway—Hemingway had the good idea of finding out what you "have to do" [as being] terribly difficult, and then the still more difficult stunt of doing it.

Of course you weren't bitchy, Xen, in not being able to put up with it. The part of society to which most of us belong has a funny sort of idealistic notion about such things. The plain fact of the matter is that there are just two types of people who can go along with deflections on the part of their spouses. Those who just simply don't care very much, and those who love so deeply, like saints, that they cherish even the imperfections.

But at the same time, an awful lot of good people, like John, do those things. Until it seems almost as tho such things, too, were part of the structure of everyone who amounts to anything. As tho it were something that has to be gone thru with, by the doer as well as the receiver.

Well hell there's not much really I can say, in a letter especially. Except that I have a sense of warmth towards you as always. And no one else could share that problem however willing. Except maybe some psychiatrist and all he could do is to try to straighten out the person involved so he'd know what he wanted to do himself. But I think it's pretty important to love and to be loved. For most people. If you can get it to do.

Your letter seems pretty sensible. Even wise. I tended to think, at first, how much easier it would be on you if you were where you could have contact with some unquestioned friend, like Tal. Then I was remembering that when Nan left me that first time, I came back deliberately to that lonesome big house, avoiding people that I knew were friendly, until after that first few days. A person in such a fix wants support, but the support he gets may not be towards the thing that's for him. When a woman leaves her husband, or vice versa, friends all say, just to be friendly and to support him (but maybe also because that's the thing they know they should do but haven't the courage) "You did the right thing." Well maybe it is, maybe it isn't; they don't know; that's just a platitude expressing a probably honest friendship. Whether or not the person ought to have done it, (but I guess in most cases like that you ought to, you have to, it's as you say, you've got to do something), and whether he shall go back, or whatever is the thing integrally true for him to do, is locked up in a person's own mind, and no one else except a genius can hope to know. Yet a thousand and one people will always say they know. All people can really do is to try to find some way of expressing the warmth of their friendship—which is what I'm trying to do. And not so well. I know. If you were here it would be anyway a little better. There is some wine in the kitchen. And it's certainly warm in here with the cold wind outside and the waves out in back. Those things may not help solve a problem, but they don't make it any worse.

> Well Xen. Hello. Hello,
> Xenia, hello
> with love from Ed to Xenia

˷

To Remo Scardigli
April 5, 1944

Dear Remo: Fact that Cornelia's birthday is tomorrow reminded me to write you a letter air mail. Enclosed with this also you will find one official invitation, never mind for what, to the lady who accepted your proposal (of marriage no doubt, I don't know why I spell that name wrong. And who is proud to call herself your wife). Who says that she cannot come down to Carmel because (a) we don't live in Carmel, (b) we don't live in Ipswich, (c) Ed Jr cannot speak Hindustani, (d) we live too far from Carmel, (e) Carmel is too far from England—she means actually Ipswich, and (d) "to one who for a weary space has lain, lull'd by the song of Circe and her wine." I felt even concerned because you should feel that she felt she had to have an official invitation in the first place. The sum of this letter is, then, that we suggest, both of us, both Toni and I, that you stop thinking, actually dwelling on the blonde Ipswich—Ipswich on the blonde, the long legged Ipswich, and you will realize about official invitations. Toni just got back from SF—took Kay up to Burlingame for Easter week—and stayed overnite with Virginia. I hope innocently. Or at least she says Virginia. State of mind. Says she actually slept in [the] same bed. These gals. What they won't get into. So anyway, Ed Jr is still annoying the boy friends of the Australian girls, I am thinking of cleaning this typewriter but never get around to actually doing it, and I had a letter from Jim. You probably heard about Jon's condition; nothing to do with England, caused no doubt by "more semen for the navy" (Toni's favorite expression, we actually hear it over the radio!). Tal is about to have her 2nd, and Ritch has been called, examined, signed, sealed, and delivered to the army. After 21 plus 10 days. They will have a bad time I fear, altho with care she may be able to stick it out in their present house. Only trouble is it's way out towards the lighthouse, and she cannot drive. Bad place for it to get noised around that a woman is living alone. Of course most of the young men aren't running loose here the way I understand they are in England, what with triplets and all sorts of scandals resulting. As I hear the story she said pursing her lips priggishly. But there are always plenty of creeps, they are not acceptable in the army and they're always the ones that cause the trouble as prowlers. I hope you follow me Remo, I'd hate to be way out here all by myself. Never like to drink alone.

It is curious tho, the way everyone seems to be flocking back here to the Peninsula. Bet Jn will be here within another year. And would be more if it weren't for the president's greetings. Incidentally, Francis Whittaker is 1A. They're really reaching out after them now. Nearly as I can see this newspaper talk that those under 26 are likely to be taken is just part of a war of nerves against married older men. Ritchie is 36, Francis is nearly 38, I am 47 next month. If he were living my father would be, let's see, is it 72. Grandfather would be correspondingly older. Altho why I don't know. If a gal could just keep around 21, I'd say that would be better. And a man; oh roughly 26. Which reminds me of that telephone pole joke. Remind me to get Toni to tell it. Well Remo look, if this thing goes on I don't know where we're going to get, and anyway a sample of oil just came in. So I get busy.

To Edward Ricketts Jr.
Monterey
April 27, 1944

Dear Ed:
Been having so much trouble with this typewriter, good stunt to try it out on a letter. Lab typewriter at Cal Pack really in pretty sad

shape. Don't suppose they can get another. Oiling and cleaning the type did help however.

We had your letter mentioning deLaubenfels; enigmatic at the time, but understandable in the light of more recent events. If you say you're "offen him," or closer than ever, we'll understand.

Such a case, you'll get to see some really primitive tropics. If you do get up there, be sure to follow all anti-mosquito etc suggestions to the letter. Not only for malaria which is bad enough, for filariasis which some of the returning hospital cases seem to have picked up probably in the Solomons. Entirely an insect born disease, and as I recall, by mosquitoes, but not by Anopheles the malarial mosquito. Good stunt to be smeared with repellant grease whenever you have to be exposed.

Last I heard, Jn wasn't figuring on coming back here to Cal until after the war. His eventual home I anticipate will be up Carmel Valley. Probably get a little ranch. But I expect that in any case he will spend a good deal of time in NY.

I have been working on and off on the Mandated Islands, about which I now know quite a bit. The Japanese scientists are great publishers, much of their work is in English, especially biolog. work, and lots of it can be found here. Also having a lot of specialists' papers bound, after I have laboriously collected them. Out of the 115-odd papers so far published on Pacific coast polychaete worms, I have succeeded in getting 100. Putting in more time than my own use and my own lifetime will justify, but will will my scient. libr. to Hopkins Marine Station, and the considerable work and knowledge I put into collecting, organizing and having them bound will benefit more than I. But in the meantime I will be amassing pretty complete data for the definitive manual. After the North Pacific trip. Can cover all the important marine invertebrates from Bering Sea to Panama. Big job. One trouble I'm having now, bibliographically, is getting ahold of papers published in Europe. Oh oh here goes our ribbon feed again. England is no problem, but no money can be sent into Sweden, and no packages can come out. And of course France, Germany, etc are impossible. Have written American consul in Portugal, but don't suppose much will come out of it during these war times.

Ritchie goes in the army last of next month. Fortunately leave him time to be with Tal and to see the new baby. Which Tal calls "little sister" but most of us call "the twins." Last time Crl swore she was knitting a sweater with holes for two heads; others spoke of kids with supernumerary legs, but Tal wasn't fazed, so nothing much we can do now. Francis Whittaker has been classed 1A, expects to be called up for his physical soon. I imagine he won't take to the army too kindly; has been his own boss for too long. Most of the people being sent in from here are pretty old. Can't understand all this newspaper talk about under 26.

Now I am familiar enough with all the work likely to pop up here so that I feel quite at home. Can put thru most of the analyses very rapidly. Weigh to the 4th decimal place on these swank Becker Chainomatics in a matter of seconds; nearly as fast as Norm. Maybe quite as fast in some cases. As time goes on, the analyses become more and more "tight," tolerances are peeled down. Now we normally shake out all the saponifieds twice, that is we extract the first extract again; all of it has to be done out of daylight, and with high degree of accuracy. So we're really splitting the last percent. Matter of fact (tho I'd better knock on wood) I'm getting fast enough and accurate so that at this season of the year, if things don't get any busier, I think I could do the routine—excluding research—with time to spare for my biological letter writing, listing, etc.

Cornelia showed up the other day. Just in time for a wonderful champagne party. Toby Street's. We ruined a whole case of champagne. He also has half case of brandy to start and I don't remember seeing any full bottle by the time we went home. But then, maybe I couldn't have seen them if there'd have been rows of them. Nancy and Cornelia did very well for themselves. Managed to have a wonderful time, sang all over the place, drink and yet not get plastered. I have a sneaking suspicion that the truant officer is only a few jumps behind Cornelia; she hasn't gone to school in nearly a year. Course one trouble is they've moved so often. Suppose she has no morale for school. She says as soon as she and her mother have settled in Laguna, soon's Nan gets back from selling the Bremerton house, she'll start in again there. But I'm afraid she'll be so mature all that will seem like kindergarten to her. Well, I suppose it's her problem, but I'd certainly like to see her finish high school.

Went to Kay's open house last night at New Monterey school. Very nice. Wonderful thing how the modern teachers manage to scrape up some artistic ability in everyone of their students. Wish it had been that way when I went to school. Kay is picking up fine. Getting better all the time. Can eat as much meat now as she wants, and several eggs a week. Toni still at the *Herald;* sort of an old timer there now.

Well I may as well stop fighting this typewriter and call this a letter.
                                        Love from all

To Xenia Cage
August 1944

Dear Xen:
Toni has written you. I heard her [talk] about it. But whether or not she mailed the letter is another and other problem. In fact she may not have. Her mother never told her about those things. Usually with that flutterbrain T it's sufficient if she even thinks a letter, let alone write it. So Kay has just learned to light the gas and there's great rejoicing. Happened just this minute. Reminds me of the time Toby (dog) learned to go down our back stairs. So excited he kept on doing it. Wanted someone out there to watch him. Like the horse "But just think, I can talk; and I'm a horse."

So we're all horses. And Jn Sbeck is here again now. The greatest horse of them all. And I'm further behindhand than ever with all my PBL work and with my correspondence. With everything except Cal Pack work and listening to the phonogr. And of course drinking. It goes at *its* merry pace. Not ours god knows. Toni cooked up the swankiest dinner you ever did see for Jn Gwen Ritch and Tal and all. Lamb, or was it veal, rice, curry, chutney and all sorts of hors d'oeuvres. And the night before I went to Wing Chongs for simple purpose of buying quart of milk. I had in mind only making [a] cup of chocolate because I was tired from work and it was nearly 6 PM. But instead there was a sign up (that store!), fresh oysters. So I switched to beer. And chopped the (very large) oysters up into cocktails with catsup (guy in the cannery steals it for us, they have it in 5 gallon cans, tomato sauce they call it, for sardines, and with oysters about!). And chopped celery, and onions, and garlic salt, and a dash of Louisiana pepper sauce. And lots of fragrant lime juice and bits of peel. You can see how things go on here.

About all that gets done is listening to music. There are many recordings of the "Art of the Fugue." As I anticipated I would find if I looked [it] up. 3 organ performances, Schweitzer, Wenreich or Weinrich on the Princeton Univ. old-model organ. And the one I have, Biggs on the Baroque organ in the Germanic Museum at Harvard Univ. Most modern organs make my tired ache. I would like to hear in actual recital that one at Harvard. Those people have everything. A library that places 4th or 5th in the whole world. Exceeded in number of volumes only by Brit. Mus., Natl. Libr. at Paris, and Libr. of Congress. The greatest university library in the world. And the two finest organs in America. What a place!

I haven't yet got the Buxtehude album, but I'm working on it. With two or three other things not very accessible here. But I have got some Bach organ sonatas only recently recorded. And lots of Handel. Seem to be having a Handel jag right now. As I did with Beethoven when I was in the army. Also have been revising my music chart. This sort of stuff: Guess you didn't see the old one, no room for it on the walls. This one has lines in different colors to indicate the nationality of the composer. Now when I do it in greater detail I find that the discontinuity of the productive times is still more obvious. Until I graphed that information for the first time, I had supposed that the flow of genius was continuous; that the old duffer handed his torch onto the younger. But it seems not to be that way at all. There are spurts, everyone stimulated everyone else, then the thing dies out, there's a period of sterility, and a new thing pops up. Just like groups of cars along a road at night. Found them on time in 6s or 7s, with occasional stragglers or hurriers in between. Or like the bumps in a corduroy road, fairly symmetrical, road doesn't wear down evenly at all. If a music lover could have his choice, he would live in the 1550–1600 time of masses and madrigals, Palestrina, de Lasus, Victoria, Byrd, Monteverdi, and the English madrigalists. Or from 1700–1750. Purcell had just died an early death, Couperin, Bach, Scarlatti younger, Handel, Rameau were all producing, sonatas and concertos. And then with the Haydn, Mozart, Beethoven flurry into the 1800s it was all over. Music was done. And the only legitimate place you can ever apply the word stinking to is to 19th cent. music. Exception of Brahms.

Or in the very early times, Dufay and the medievalists. 1400–1450, only des Pres bridging the gap to Palestrina. Before that plenty of music, but anonymous. And except for Gregorian, apparently unintelligible to the modern ear.

Well I suppose plenty of people know all about those gaps and fullnesses, probably studied in courses in music history, which thank heavens I was never exposed to. More fun to work it out yourself. Dear Xenia, the letter continued, may I communicate to you the result of my researches in varied fields? F as in food. And Alaska. Toni wants me to make her one of those charts for painting. And I probably will. Revise my old one. A as in art. And Pebble Beach. Jenkins people were over here once since that fine occasion. You probably forgot about all those bottles of scotch, and the martinis, and the beer, and the fine Portuguese wine. Going over there tonight for dinner. Certainly something to know an associated farmer. Seems to be a pretty good guy.

All of which seems to be getting me in a mood for cooking a masterpiece of chopped bacon onion and curry into which some eggs get introduced. And eaten. Before I famish. If this finds you in the midst of work at the office, 9 to 5, half hour for lunch, I hope you will have the gumption to go out and get you a sandwich specially made. Someone ought to put offices where they belong.

> The music chart Ricketts mentions was most likely a replacement of a similar chart lost during the lab fire of 1936. His sister Frances remembers that "he made a graph depicting music as paralleled by other arts, discoveries and world events, which went around the wall of the lab, and which was destroyed by the fire" (Ricketts E-mail, September 5, 2000).

To Joseph Campbell
September 23, 1944

Dear Joe:
The time of Xen's visit here was very full and very pleasant. Left not much leisure for letters. Now that she's en route back there I'll put in some office time.

Very grateful to you for sending on the new book. I have dipped into it only the least bit, but already from that little I am confirmed in what I would have assumed sight unseen. Certainly a front line, firing line of a most important issue, more so because not well known popularly. The one that underlies all others. However did you manage to get so much lively interest in what I would have expected to be a rather dry guide!

We got your Navajo portfolio. Wonderful reproductions! Xenia said you were disappointed in them. Have to see the originals to understand that. I was badly disappointed in the *Sea of Cortez* color illustrations. In the black and white too, for that matter.

Cannery Row about the same. Busier than ever. The lovely nice sporting house, Flora Woods, was a war casualty. Too much gold star mothers, and slightly unreal attitude of Army Med. Corps (which I was in for about a year as Xen no doubt mentioned; 2nd time in the army for me, 1st world war too). Toni has been working on [the] local paper, but recently quit. Doesn't want full time work. In any case we keep too busy. But still a little beer gets drunk, so I suppose everything is alright.

I have been working on my little animals, on and off. An ultimate manual on the Pacific Coast marine invertebrates. Which has to be preceded by more exploration. Chiefly a trip to Bering Sea, Aleutians, Gulf of Alaska maybe over to Kamchatka. And lord only knows when I can get up there. But I can work on the prerequisites. Also lots of dope on a projected popular book on the Japanese Mandated Islands, on which, by devious ways I got to be something of an expert. But that's probably out now, with military turn of events. And I would have finished it in any case only if I got publisher's contract, which seems not to be forthcoming.

John is publishing book on Cannery Row which should retire me, in self defense, from the public eye for a few weeks after publication. Aside from that phase, and excluding a few sentimentalities, it could be a good book. Tortilla-flatish.

In the meantime I have been chemisting for a past year. Into that directly from the army where I was doing clinical laboratory work. All of which made hash of my own programme but the threads are still there and maybe I can pick things up again.

My best to Jean.

I'm assuming that you know we'd like very well to see you people again.

> The "dry guide" Ricketts mentions is most likely Campbell's *A Skeleton's Key to Finnegans Wake*, which was published in 1944. Another major project Campbell worked on at this time was commentary for *Where the Two Came to Their Father: A Navajo War Ceremonial*. It contained text and paintings of a Navajo ritual in which warriors prepared for battle and death. The book was published in 1943.
>
> Upon his return from Europe and Africa, where he had been

working as a correspondent for the *New York Herald Tribune,* Steinbeck composed *Cannery Row.* While soon after its publication Ricketts was thrust into the public eye, his letters reveal his patient, good-natured attitude toward Steinbeck's intentions and the subsequent disruptions in his life due to *Cannery Row*'s success.

To Edward F. Ricketts Jr.
November 15, 1944

Cal Pack 11-15-44

Dear Ed:
You can regard it as a belated birthday remembrance, or however: couple of months or so back I started subscr. of *Downbeat* to you. So if it starts arriving suddenly, no apparent reason, that's the answer. Suppose you may be getting it already; imagine not; if so, pass it on.

What reminded me was that I supposed I had forget [*sic*] Nancy Jane's birthday. Yesterday the 13th stuck in my mind. I wonder whose? Anyway today I called up Aunto [Frances Strong] and she said 28th.

I suppose you're still on New Guinea. Lacking information to the contrary. But who knows when you'll be in the Philippines. Soon now I suppose, altho the Japanese seem to be putting up a lulu of a scrap. Certainly won't give up until they have to. You certainly have to give them credit. As geographers and scientists as well as warriors. Watta race!

I gave up work on the Mandated Islands book. Figured I'd clear it up only if I could get contract for publication and none has been forthcoming. But I did dig up a lot of dope not commonly known, some of which I imagine the apropos branches of our government either don't know or aren't making use of. Such as detail maps. But I have conveyed most of the info, and that's all a person can do. Most interesting thing: after husbanding along the native population for years, the Japs report a 20% decrease in one year. With no sign of pestilence, famine or natural disaster. Population of one island in the Palaus, Babeldaob [Babelthuap], the largest, was at one fell swoop reduced from several thousand, I think 5,000 to couple of hundred—that is the native population. At the same time the Jap population climbed. That was only a few years after Jap mineralogists found aluminum ore on the island—bauxite. I'm still wondering what they did with all those natives they had been nursing along so many years. Probably never know.

Jn bought the de Soto adobe and has established himself permanently in Monterey. In the meantime was living at Carmel Highlands. But what with gas rationing on the part of him and his friends, and Gwen not driving, it was inconvenient. The baby is fine. They have a nurse to look after it, but Gwen is a pretty competent mother. And a good cook and housekeeper. They maintain a small establishment, no servants other than the nurse. Jn asked T to do some typing, and I imagine she'll get back pretty much to be his [secretary].

Cannery Row still set for release right after Christmas. I haven't seen any more of it tho still expecting a set of page proofs. *Life* in the meantime put in a tentative suggestion that they photograph Cannery Row and run one of their little essays on the lab, on me, on Cannery Row etc, but I didn't approve, neither did Jn, and we turned it down with the knowledge that still, if they want to, they'll come in and photo the street, Wing Chong's, the outside of the lab etc. But I doubt if they'll do it without our connivance. If the lab and if I could be left out I wouldn't care, but that's chiefly what they want, and I dislike the publicity. Jn is in the meantime planning on Mexican movie based on the pearl and the La Paz Indian legend in *Sea of Cortez*. Or did I write you? Well. So generally busy at Cal Pack lab I've had no time for PBL work nor my own. Right now fairly idle tho.

Love

> The Steinbecks moved into the "Soto House," located near the waterfront in Monterey, in late 1944. Gwyn had given birth to their first son, Thom, on August 2, 1944.
> Although both Ricketts and Steinbeck disapproved of *Life*'s inten-

tion to send a reporter to Cannery Row, *Time* reporter Bob DeRoos and *Life* photographer Peter Stackpole did, in fact, visit. According to an interview with Martha Heasley Cox in 1975, Stackpole recalls his meeting with Ricketts. "We ran into Ed Ricketts, who was a close friend [of Steinbeck's]. Ricketts was quite friendly with us and allowed us the run of his house. We photographed him at work and down in the basement and in his lab where he was doing various experiments with sharks and various fishes—which he was doing in the laboratories. . . . He was living upstairs, and I remember he had music going all the time. He was very partial to Gregorian Chants."

Stackpole took many photographs, ultimately reporting that Steinbeck's depiction of Cannery Row and its residents was accurate (*Cox Steinbeck Newsletter* 9.1: 20–21).

∽

To Edward F. Ricketts Jr.
January 6, 1945

Mailed about
1/6/45

Dear Ed:
I am sending along this notice pronto, airmail, so that you won't try to buy there—altho where I can't imagine (Manila bookstore maybe by the time the package gets there)—a copy of *Cannery Row*.

I am sending you one under separate cover. May take a long time to get there but it's on its way. Pat Covici sent me one, and Jn had already given me one of his, so I'm sending you the copy Pat sent. Jon annotated it to you. Incidentally, hang onto it, a collector would probably rate it at fifty bucks; Ed Ricketts Jr's copy signed by Jn of a bk that concerns Ed Sr. Or so book collectors figure.

Very pleasant Xmas. And fine New Year, featuring, among other things, another of those bang-up parties at Toby's. Gallons of champagne punch, 53 people, Nancy Jane and boy friend among them.

Usual round of activities, Toni and I managed pretty much to get ourselves tired out.

Well, OK. Have fun.

To J. E. McConkie
January 18, 1945

Main Office Laboratory
Plant 158
Jan. 18, 1945

In line with Norman's suggestion, I am summarizing certain aspects of my personal investigations during the past decade.—Especially those involving the relationship between plankton production and the natural history of the sardine.

This summarizing is being done at odd moments during these otherwise slack times. It seems likely that a compilation of this sort might be valuable to CPC [California Packing Company]. Or in any case quite interesting.

Much of the material is fairly fresh in my mind. Just prior to the war, I made a revision of "Between Pacific Tides" (which should be published soon now that the paper shortage is lessening). I considered including in this second edition, a resume of the plankton work on this coast, and have been keeping fairly up to date in that connection.

Recently also I examined the stomach contents of several lots of sardines from the November landings. Norman thought it would have been good if I had had access to one of the regular CPC research notebooks for this phase of the work, and in that particular case it would have been both appropriate and convenient. But actually, very little recent lab work has been required. For one thing, fresh material to work with has been scarce. And for another, plenty of lab work has been done already on these organisms, and the need now is for analyses of the data.

So if all this is agreeable to you, when I turn in my summary, I'll merely report the Cal Pack lab data rather more fully than if I had recorded it in one of your notebooks. If not, I can copy out pretty quickly my relatively few recent notes, in whatever form you require.

Norman suggested that I write to you in this matter.

⌒

To Xenia Cage
February 9, 1945

Dear Xen: nice letter prompt answer
or none. And my desk piles higher. But not today

And from Jn in Cuernavaca, small house, large fine garden, no shoes, black skin black dirt and Gwen and the baby all black sounds fine but I can see also tropical green and Jn busying himself wondering about some new insect.

Also from Ed in Manila who just barely knows there's a war. But who yet still that far away follows *Downbeat*. Isn't it fine that this can be sent to GIs. Ed's chief apparently joy. Except Manila night clubs and perhaps Phil. girls. How I envy what they can teach him if only he'll relax and lettum, or if he won't perhaps I can consider teleportation his father. But best of all, that finally he's taking piano lessons. And now they'll do him no harm. Crazy crazy business. Ed may have been thru Leyte, he may be in Manila after Hollandia and a year in Australia, two years overseas and hard training. But mostly what he notices is that now he takes piano lessons. Well so I feel not so bad for the world.

Reese Music in the Middle Ages, VG V Vgood. But pretty scholastic and dull. Redfield "Music A Science and an Art" with data on what he calls the "Chord of Nature" in which he realizes our scale lacks the 7th, the 13th, the 17th, the 19th partials, tho it has others up to the 45th (last must be pure accident, no one has such a good ear!). But I can hear the 7th very plainly in a measured guitar string. And of course see the node. So finally I get out my tuning forks. Funny thing. C 256 and C 512 were wonderful ringing clear, you rarely have such wonderful octaves on the finest piano. So I enjoyed just listening. Until I left the forks around one night and that nice not too bright sitter came in to stay with Kay while we were out no doubt getting ourselves thoroly oiled and I wonder why you weren't there such times to miss! So anyway the next day C_512 was sour and didn't even sound good alone. Think he tapped it with a table knife, anyway one side was slightly nicked. A queer thing: C_256 followed by 448 (the 7th partial) sounds fine after a few times, but you can't go from there to C_512 unaided. Probably could with practice. But not naturally. And 256, 228, one other I cannot now identify don't know whether it's in "scale" or out, followed by G_384 and 352 (11th partial) sounds very natural; after the first two, you sing the others naturally. Seashores Psychol. of Music has also reports of the first large scale oscillograph—harmonic analysis of single actual tones. Of the violin G string for example (which gives only .1% of its fundamental, most of all the first 5th above, did you know that, I never did.). So I'm having lots of fun. Getting somewhere in understanding that crazy mix-up. I have use of Cal Pack's calculator, so can run off now in a hurry things that by long hand took me hours. I spent three hours working out backward the factor for the even-tempered scale, 1.05946 or something like that, then found it already written down! When I get back I will get tuning forks for the equal-tempered (piano) scale, perhaps in the same fundamental for which have already the diatonic. And can then compare them. But where does Redfield get the idea that to sing in the diatonic as opposed to the piano scale is singing in a "natural" scale. How can anything that lacks the natural overtones be a natural scale. Well perhaps I'll find out that one sometime too.

Probably Tal wrote you about the new singing group. For old music. The Kyrie Eleison from the Gregorian Missa de Angelis, Palestrina "adoramus te" and a Morley Madrigal. Unaccompanied. And very difficult work. Toby is happier than I have seen him. Says "tickles me pink, after these years, to discover I'm not the *only* one who gets off in singing." Well if she didn't I won't sezze [says he] stopping after the fact.

Your analysis apparently "taking." Mine didn't. I had none of the terribleness, and of course none of the curative effects. Except perhaps in the long run. Altho I cannot still to this day relax in the usual things, dancing, singing, swimming. And perhaps never will. Altho I can well enough relax in some things I consider more important. So OK.

"be in bed surely" you sound fine to me Xen. What wisdom! Wonderful, mature, to be emulated wisdom. And with whiskey. A person who can reason that way has no business with a psychiatrist. Except maybe if the psychiatrist gets mixed up. To straighten out him or her. But that's the depth of it alright. That it's alone. That it's so entirely and completely alone. That a person most of his life has to walk alone. Why not a real crisis. Of course it is. Just as real as a physical danger; that auto just missed you or an infection nearly takes over. No underlying difference. Any of them equally could be ruinous. Conscious perhaps, but certainly not intellectual. Anyhow not *only* intellectual.

Speaking of the ghastliness of champagne cocktails that have had ginger ale mixed in them, you would certainly like that Jennifer [Lovejoy]. Unbelievable. No one could believe who hasn't seen that child stalk any unattended glass of wine, whiskey, beer, rum. But beer most of all. A sip and then the sourest face you ever saw but a happy interested sour face. And then another sip. Not going to have any trouble getting that child to hold her own in a group. Curiously extroverted or so it seems. Just demanding. Not a bit shy or sensitive or retiring. Of course it might be different later on. But there's a child who will think as we do about drinking. Hope she doesn't overdo it.

Well. I continue to write and I have work to do. Toni just back from giving a talk on fascism. Which I read. A brilliant job. We are off to

Nootka (outer coast of BC [British Columbia]) if we can get transportation, [on the] 1st next month. So another least little glass of wine, and a letter of intro for T to SS company—trying to blast from them the sailing date of the outer coast Vanc. Id. steamers. And she has to go all the way to SF just to get it. Wotta business. Danger of subs still I suppose. Well have fun and do lots of good work I bet.

In April 1945, Steinbeck, Gwyn, and Thom moved to Cuernavaca, Mexico. While there, Steinbeck worked on the film version of *The Pearl*, which was eventually released in 1948.

Ricketts and Toni did not leave until May 1945 for their trip to Canada, where they collected in the region west of Vancouver Island until early that July (Hedgpeth, *Shores* 1: 44).

∽

To Edward F. Ricketts Jr.
April 19, 1945

Dear Ed:
It occurred to me that you, prowling around in 2nd hand book stores in Manila—which no doubt there are and no doubt you will—might pick up some things I want very much. Manila is a particularly likely place because these several things are about the Philippines and it's just barely possible, what with upset conditions and all, you might find them there, and they might be within reason.

Any of the Semper (Carl), Reisen im archipel. Philippinen (Carl Semper's *Journeys in the Philippines*) also as "Reisen im Phil. Arch." Are very desirable, but particularly the sections on holothurians and marine shells. The complete set is something like 10 big quarto volumes. It was published in Leipzig Germany, I forget by whom, from about 1867 to the early 1900s. The complete 10 volumes are excessively rare. Univ. Calif. library doesn't have it. One of the US 2nd dealers in scientific periodicals recently offered a complete set for something like three or four thousand dollars. But Fisher at the marine station has the volume on holothurians (sea cucumbers) and he says he picked it up in England for only a few dollars. So if you run into any of them, keep all this in mind.

Another Philippine thing that I have been trying to pick up (because it transcends that local fauna and becomes a work of importance on tropical zoology in general) is Grube's "Annulata Semperiana" published in one large volume with many fine plates in the 1870 I believe. Have St Petersbourgh Academy of Sciences. I forget the volume and number but if you run onto it you'll recognize it by the title. This also describes some of the Semper collections.

Carl Semper was a German zoologist, a very great man, who with his wife stayed several years in the Philippine and Palau Islands collecting (mostly) marine specimens in the 1860s, and studying the coral reefs. His wife painted the living animals very accurately and many of her paintings were reproduced as color lithos in the scientific reports. The bulk of the work was published in the Reisen above mentioned, but a few of the groups were described separately elsewhere. Interesting because they are so complete and accurate. In many cases there has been nothing since published on tropical zoology that supersedes them. So, grab onto them if you can.

Cannery Row is coming up as a movie and when the film is released I'll just have to move acct publicity. But so far it hasn't been bad. Most of the people I've met as a result of the notoriety have been very decent. But of course the film will send all kinds scurrying to Monterey to look up Doc. Still hoping to get up north this summer, but the fact that I have had so little time to plan the trip, and the possible difficulties of transportation incident to the SF world conference are factors working against it. Jonny boy is back in Mexico, working on the Pearl of La Paz episode as a movie. Hilary Belloc has been down. Barb and Ellwood are back here as perhaps I mentioned. Well, this was just a grabbed few minutes at Cal Pack, where I have been mighty busy for couple of weeks now. Write when you can.

Originally, a film version of *Cannery Row* was to be made with Lewis Milestone and Burgess Meredith; however, the project was halted because of a lawsuit filed by Bernie Byrens for breach of contract in which he alleged that he had been given the rights to the film (Benson 610). A film was eventually produced in 1982.

&

To Editor, *Monterey Peninsula Herald*
April 26, 1945

Editor,
Peninsula Herald,
Monterey, Calif.

Sir:
The anonymous ad in Monday's *Herald* entitled "Organization to Discourage Return of Japanese to the Pacific Coast" re-emphasizes to me Hitler's essential success as a teacher.

We think of this former German leader, with his theory of racial superiority, as a complete failure. And so he is in his own country— a military failure at least, soon to be discredited and forsaken.

But in this asylum of the oppressed, his seed bears fruit. It is ironical that only here his ideas live on, in a country built on a theory he shunned—the theory that all men, created equal, must be allowed equal opportunities in life, liberty and the pursuit of happiness, regardless of race or creed. He lives on in the minds and hearts of brave Americans (hoodlums, some will call them) who, despite great personal risk, fire from speeding cars into the homes of fellow Americans. Or, better, who burn the property of fellow humans whom they condemn sight unseen because of race. And here also thousands of equally brave guardians of democracy are patiently erecting a structure of prejudice on which the war of the future can be based. A real war this time—a credit to his teachings!

In due time, these ideas of racism can be applied successfully to other recent Americans, to Filipinos and Negroes and Jews, to children of Chinese, German and Polish ancestry. Until finally there shall be left here (a unique country!) only the few thousand yet remaining of our true original natives, the American Indians.

And so Hitler, suffering in his own land a death perhaps neglected and dishonored, richly succeeds here in America. Truly a prophet hath no honor in his own country.

Respectfully,
Edward F. R. Ricketts

1946–1948

To Department of Oceanography
University of Washington
Jan. 8, 1946

Dept. of Oceanography
University of Washington,
Seattle, Wash.

Gentlemen:
I am wondering if you have any serial records of volumetric *total* plankton other than those reported in Thompson, Johnson and Todd 1928, Johnson and Thompson 1929, and Thompson and Johnson 1930.

Probably the fact that nothing of the sort has been published indicates that no additional observations have been made. However it seems possible that figures for the remainder of 1929 might be at hand, to complete two consecutive calendar years.

A revision of "Between Pacific Tides" is in the offing. While I cannot be sure that a resume of plankton work on this coast would be publishable there, or even suitable, still I have become interested personally in liaison work of this sort. It might be fun to attempt to diatom production figures—up there and in southern California. Perhaps this has been done already.

With the end of the war, and the end (we can hope!) of the paper shortage, I presume that publication of the phytoplankton calendar mentioned in Phifer and Thompson 1937 will be forthcoming. Or has it already appeared and I've missed it?

       Sincerely,
       E. F. Ricketts

Bob Gibbon was collector of customs and commissioner of immigration for the navy stationed in San Francisco during the war. In a letter dated January 15, 1946, he wrote the following to Ricketts: "I hope you remember me. [. . .] Libby Cass brought me over for several very pleasant wine and record evenings while I was twiddling my thumbs as CASA. [. . .] Have been told shark-fishing has been profitable off the Calif. coast, primarily for the liver and the vitamin(s) therein. [. . .] You are the only person I know who would [be able to] give me the dope."

To Bob Gibbon
January 25, 1946

Dear Bob:
It happens I do know something of the shark situation. And glad to tell.

Livers of some sharks are among the most valuable of all sources of vitamin A. Others have very little value. It depends mostly on the species, secondly on the sex ([males] are more potent but any of us could have guessed that!), then on age and lastly on condition. Old male soupfin shark livers are best of all. Up to twelve and fifteen dollars per pound at one time were paid for such livers and at first the fishermen had literally [a] treasure trove. Now the soupfin on this coast has been depleted to the point where little is done with it except incidentally by the commercial fishermen, and new fields are being sought to conquer. At the same time the price ceilings have been depressed, perhaps with the government's decreasing vitamin requirements.

Still however there is something left in the way of a market, and somewhere there must be adequate supplies to fill it. My unconfirmed impression is that not many of the tropical sharks have high vitamin content. And many of them I know are too low to bother with.

A pamphlet which has just been issued which you could recommend to them down there is

"Guide to commercial shark fishing in the Caribbean area." Fishery leaflet No. 135, Fish and Wildlife Service, Wash. D.C. 1945. 149 pp illustr.

On second thought, never mind; I will write from here asking them to mail it to you direct. Save at least a couple of weeks.

Specifically: Only the liver is valuable. The vitamin content of the viscera has been much discussed. It has some A. But so little, it's so hard to extract and goes bad so quickly it doesn't pay to fool with it. And to 2) and 3), both yes. Per below.

Take samples of the liver of a given species (but don't mix up samples from more than the one species), chop them up, put them in a kettle, bring to 180º F for a short time (heating depletes Vit A) long enough only to sterilize and extract the oil. Draw it off directly into a clean screw-cap can (8 oz. sample is fine, even smaller will do), keep out of the air and light, send at once with letter to:

> Percy H. Fish, Head, Vitamin Oil Dept.,
> California Packing Corporation
> 101 California St., San Francisco 19, Calif.

This is not a good or efficient method but it will get an index of the richness of the species found there. Send a separate sample for each species. Also bear in mind that sunshine (any light for that matter,) destroys vitamin A. There are methods for sun extraction of liver oil, but the vitamin loss is considerable.

If you find that any common local shark has high vitamin count, shipping methods can be worked out. We get bonito livers very frequently from Peru, and in the past year we assayed shark livers from Uruguay and Brazil, and cod from Greenland.

If I were there, I'd be charmed and interested to have a look at the marine animals, but most of them are ephemeral and require highly specialized collecting methods. However if you yourself are interested, I'd like to see representatives (in duplicate or triplicate, since determination usually requires that one specimen be sent to the specialist in that particular group) of the commonest shells and corals, dried. It adds a lot of interest if they have been seen alive, so that something can be known of their comparative abundance, where and in what animal societies they live, etc. The chief variables in animal distribution seem to be (a) temperature (which doesn't operate within a region and can therefore be disregarded altho it might be a factor in comparing animals from the northernmost and southernmost Marianas), (b) depth or position in the tidal level (some things occur high and some low in the intertidal), (c) protection from or exposure to wave shock, (d) type of bottom (sand, mud, rock, coral etc). I'd like eventually to get a knowledge of the commonest shore invertebrates, the "horizon markers," of the waters adjacent to the Pacific coast of [northern] and southern South America. Of course that's an impossible ideal, but I'm glad to make steps in that direction whenever possible.

When you get back, stop in for more records and wine drinking. And talking. Which seems to go on here always. Always and pleasantly. This typewriter is not always a pride and joy.

I hope the shark livers turn out to be valuable to those good people who must have been sadly upset by the war. It will be hard for them to go back to pre-war standards after so much excitement.

                      Sincerely,
                      Ed

> The three questions Ricketts answers for Gibbon are from Gibbon's letter of January 15, 1946: "Specifically, the questions seem to be: 1) Is the shark and/or his liver or other guts valuable?; 2) If so, is there a simple preliminary process which could be done here to permit economical shipping (refrigeration seems out of the question, at present anyway)?; and 3) Would any US chemical or canning company be interested?"

To Nancy Jane Ricketts
March 28, 1946

Dear Nancy Jane:
Children is good and of course grandchildren is fine; I suppose you are very excited. But apparently having a pretty good time of it or you'd have been forced to quit work before this. Ed seemed to take it all in good stride, Toni was quite pleased but Kay was very excited.

I suppose Ed does in fact write letters. In fact I know he does, because he wrote one since he got back. Just yesterday, to his pal Parfit. And before that, to the rationing [board]. But to get him to write another might be going too far. Of course you might work on him yourself. By letter. And then who'll get you to write the letter to persuade him to write. Said the old woman in the shoe. Who beat the pig over the poke.

I am relaxing on 4 hours a day four days a week now as perhaps I mentioned. And Toni is working like a perfect slave just as many hours per day, but *five* days per week! So that leaves me time for PBL work which at the moment is certainly building up. Orders coming in, at this slack season, almost like fall. But of course the whole thing doesn't amount to much. Badly run down during the war. Schools certainly are needing things. And don't concern themselves much about price. Curiously also, some fine pelagic stuff has been coming in. Four or five of us including your illustrious brother, got more than 2,500 Pleurobrachia in a few days. Only the 2nd or third time they've come in [greater] quantity for twenty years. The stuff we used to get at Allyn. Remember? And other things have been coming in here, including some rare items not seen locally since 1924. Since you were an egg.

Ed very busy on photographing. And trumpeting. And sleeping late. Eating much. Playing the phongr. He's having a fine portable made for himself. Turned in his old phonogr-radio. Got seventy five bucks for it. Almost twice what he paid for it five or six years ago.

Also he's been making some charts of plankton for me, then photographing them. And of temperatures. Generally making himself useful. We went out last night and got a couple of hundred snails.

∽

To John Steinbeck
Pacific Grove, California
April 9, 1946

Dear Jn: Yr. letter of Apr. 5th. What a luxury that you have a typewriter, and that you use it for writing letter purposes, and best of all that the rubber stamp remembers your address because I never do. Another thing I wish is that fishermen who come in the lab to watch me pack frogs would for Christ sake stop talking about the funny guy at MIT who taught them analytical geometry and who always wandered off the subject, "always talking about binomial surds. Say Doc do you do anything with forminifera, I got quite interested in them once; connection of invertebrate paleontology."

Wandering on "Little bit drunk tonight; don't mind me; my sister and I thought we'd get out of this, so I went to MIT but hell I guess I got the lowest grade in English—require it there you know—and so here I am fishing again. But she's chemist now for Carnation Milk people, doing alright." But the worst of all was the gardener who strayed in one day, watched me fooling around with alcyonaria, recommended that I go to the Andaman Islands, acct of the vivid display of invertebrates there. Stay out of the lab, best thing to do, walk the streets, won't run up against things like that.

T. opened chg. account at Holman's (urging me to use it for whatever small purchases I make—which I won't damn 'em) because with the monthly statement they enclose certificate entitling bearer to purchase nylon hosiery when and if. But they didn't to her. Under false pretenses I call it. We had last Sunday the only really bad *Cannery Row* experience to date. It was pleasant noon, I was sitting barefoot and with nothing on otherwise but [a] shirt drinking coffee, reading. Toni just got up and was reaching for her dressing gown. Kay going out had left the door unlocked, which we're careful not to do Sat. or Sunday; there are tourists. So this god dam fellow came in saying not a word, started to walk in our bedroom. I said how do you do [and] blocking the way asked him if he was looking for someone. Toni got behind the doorway. Said he wanted to see Doc, that his wife dared him to come in. With some actual pushing I headed him out into the office, closed our door and he still had the nerve—tho I told him now this was pretty much of a private dwelling—to say he'd like to bring his wife in. At which here she was coming up the stairs. Ended up that I actually pushed him out the door against his wife, shut and locked it in his face with as much nonviolence as I could. Well I didn't get angry, never did feel even actually unkind. But it proves something or other alright. His last words were "wait a minute, won't you tell me what you did with all those frogs, those tom cats; bothered by a lot of people coming in you ought to charge admission." The others have been good people—very much moved by their own inward most kind projections onto what they think I am. And they certainly merit gentleness. But this guy was what you call unsavory at least.

What'll happen actually is Hearst will get a picture of your house, blow it all up, golden staircases that sort of stuff. At the house warming the elite of wealth and beauty; special models of B-29s operated by ten thousand dollar mannikins [*sic*] will scatter specially made perfume on the specially made (by the guests) Power's models; beggars, union leaders, small business men stand outside in the snow.

Funny thing about that beating business. If you want to make a saint out of an ordinary man, all you have to do is to beat him hard enough, long enough and often enough. Must be everyone's potentially a hero. Gal who was our biochem. asst., just went up to main office lab acct no work here, came back from Korea just before the war where her uncle was prof. in Keio University. Perhaps I told you this; no matter, you perhaps forget as fast and good as I. Japs found in his carbon copies, in his letters. He types like I do. And of course it was code. So they beat him and gave him the water treatment, very conscientiously. Blanche says "funny thing, he isn't a bit sore. No bitterness against the Japs. Against anyone. Just as kind and mellow; you'd never dream he went thru such a thing."

Probably told you about that fabulous fish. King of the Salmon. Came in with the pelagic stuff. Louisa Jenkins going to paint it. I had in the meantime given it to Rolf, but he's glad, says go ahead. This cannot be released until the 15th publicly; Rolf got his Guggenheim. $3000 for one year to travel to the museums of Europe, Asia, Australia, examine type specimens of lantern fish. Straighten out the whole group. Guess he got a sabbatical from Stanford coincidentally also, otherwise couldn't afford it; this way he takes his wife as secry. Leningrad to see Schmitt and Adriashev. Calcutta. Of course London, Copenhagen, Berlin, Sweden. I will have him get me Sven Ekman: Tiergeographie; can't myself contact Leipsiz and can't find anyone who knows a GI there. When I apply for a Guggenheim to straighten out the Gulf of Alaska, Bering Sea, Aleutian, Kamchatka situation I bet I won't get it so easy. I bet I won't get it. Sudden rush of specimen orders, in part because of the wonderful pelagic haul. But not only that. So I've been busy there. And at Cal Pack too now that Norm is gone and I'm in charge and on part time—16 hours per week, very pretty very very pretty. So no time for finishing up the plankton paper. In the meantime tho I've got the 1940 sardine paper by the Japanese planktonologist, on growth, races, migration etc. They certainly know more about theirs than we do ours. Which surprises me not at all. The English are undertaking a huge fishery investigation programme thru the Univ. of Hull. Mostly regarding the herring. Probably what you were mentioning in a recent letter. All British herring boats now are equipped with a plankton recorder. Wonderful device with a 30x lens automatically focused to show up the herring food (copepods) and the boats fish wherever the device records suitable conditions. Also the steamers operating from England thru the N Sea now are equipped with automatic plankton recorders. I saw some of the papers. If they didn't cost 10 shillings each I'd buy all of them. I guess the Nan thing may be pretty bad. Sleeping dog doesn't lie forever. Fortunately I have practically nothing. I guess Ed coming back and Nancy Jane getting married and going back east reminded her that

I needed a sail or two taken down. Or maybe she's spent all the money she got from the sale of those houses and really needs it and thinks she can get [it] from me. Peter Ferrante's handling. My suggestion but I talked it over with Toby because he's the logical one and very friendly, and he approved. That end of the paper.

> Peter Ferrante was an attorney for the Monterey County Trust and Savings Bank. He was one of the most trusted men on the Monterey waterfront. In an article she wrote in the September 1947 edition of *What's Doing*, Toni describes Ferrante as a man "who can mediate, or arbitrate between all the elements of the fishing fleet itself and who can stand between the Sicilian and the Anglo-Saxon worlds, understanding both and explaining each to the other" (24).

To John Steinbeck
April 14, 1946

Dear Jn:
I haven't gone over the past records thoroly yet, but near as I can tell from memory, this *is* the year of the 19.6 yr. tidal cycle peak. Which may or may not be related to all the fabulous pelagic phenomena, and no sardines, and shifting of water weight earthquakes. And published reports that red water in south Calif. (dinoflagellates, but this time apparently the non-poisonous type) was so intense that many shore invertebrates including even lobsters, were killed by suffocation—low O_2 content. Dick Dettering just back from Prince Rupert and Anchorage says that up there the red water was so fierce that crabs died, and it was forbidden to sell them for food.

I didn't put it all together until I started to look up June/July tides in connection with the Queen Charlotte trip. There was a minus 2.0′ tide staring me in the face for SF, June 1st. I never yet collected here on a tide scheduled so low, tho I understand that once a −1.9′ scheduled at SF fell actually to −2.3 on the staff. Due to a heavy wind from the land.

Ought to be more work on cycles. Lemmings, natural production of wheat, economics, purchases of furs from trappers. But it's proven so far a thankless and frustrating task even tho fascinating. Sudden plankton imbalances. Couple more lifetimes needed. Good long ones too.

When I drove to the [post office] there was an old old woman walking along the street eating an ice cream cone. Perhaps took all her life to achieve such freedom. Probably [the] widow of a minister, intolerant, or power driven or just ineffectual and now dead a long time so the influence is gradually fading. Hearing Ed's phonogr. (new one, Dick built it) reminds me that he has a lovely life. A personal thing of course tho, just for him only; wouldn't suit me, or you. His records are right beside the bed above the phonogr. First thing in the morning (morning being noon) he lies there up to several hours and just plays jazz. And last thing at night. Doesn't have to turn over. Now he makes his bed every day. The army was a good thing. Expect to have some news for you next few days about the lab. This year's spring name is already preempted, but I have a beauty for next year: "Rose-Spit-Gas-and-Whistle-Buoy" (p. 269, sailing directions for British Columbia, Vol. II). Now Gwen has six weeks, I bet she's counting each day twice. Tell her it'll happen as I come to the end of this page, almost before she knows it.

After a difficult pregnancy, Gwyn Steinbeck gave birth to John Steinbeck IV on June 12, 1946.

To John Steinbeck
August 12, 1946

Dear Jn, I guess you'll have to bear with my tirade on use and nostalgia. A common word with you, one of the few who know its power. Incidentally, did I send Haas' list of mollusca to you, in exchange for sending him the letter addressed to you? If I didn't, send it on to him if you will. It relates to a whole gang of Charlotte Ids. specimens I sent him. Kay is back in the hospital. Same thing, but this time recognized as not malignant: said still to be probably mortal eventually. But the operation set for Thursday has only 30% fatality. Against the astronomical prognostication before. So I feel pretty confident. Fender, Stanford University Sch Med.

But the turn of events of another sort got me to considering the relation of use and tradition. Only result being a sense of frustration. Allegedly, man is the only logical animal. His illogicality amazes me.

How could there be more illogicality. We can't buy bedsheets. Gradually we get used to sleeping on blankets. The manufacturer's strike is easy to understand.—But if it were a workers' strike that kept us sheetless, wouldn't the newspapers howl?—The amount of cotton needed for a two or three dollar sheet, dyed and designed and advertised as a swank woman's dress, brings $20 or $50 or $89.85. The people may not be served, but profits are!

But that's both understandable and logical actually. Right at the moment I'm frustrated by a pica which is both illogical and profit consuming. A tribute to tradition only. Like our calendar, like our system of weights and measures.

I can't get forms printed up for accurate use with the typewriter. Along about the 5th or 6th inch, my typing comes halfway between the printed lines. The pica, the printer's unit, is just slightly less than 1/72nd of an inch, and hence incommensurable with the inch which is the basis of typewriter line spacers. This has been going on officially since 1880, so long we've lost track of it. We don't even think about it anymore. We think of one in terms of another until we're brought up short by the incontrovertible fact, and I for one get a fine feeling of frustration out of it. The printer assures me that he'll adapt my printed form for use with the typewriter. Then he doesn't. He can't. Why didn't someone tell me this!

Why doesn't someone say to me right nowseehereeddiethis humanis prettymuchofafoollogically [*sic*]. His profit system, eating itself, eventually deprives him of profits, even of the actual dollar profits. Power gets mixed into it. Or conservatism more often gets involved. And even profits aren't served!

The things that served our youth, our childhood, maybe even our infancy, are the really important things. Around these our most primitive nostalgic emotions get entangled. And so we want to preserve these things at any cost, not for the things themselves, but for the associated emotions. Because on a certain day in the morning of my life everything was well with me. The straw roofed house and the castle in the distance and my flaxen haired mother throwing the slops out the door got related to that wonderful feeling of security and well being we get when we're young. So when I grow up, I want my children to see that same castle on the hill. No one can talk me into any new fangled notion of hygiene. Or art. Or government. When I was seven years old, that sunny morning was so wonderful, that I will have it continue for my children and their children. And any change will be over my dead body. The noble says "These people don't want to change, they like to be serfs, and poor, and plague ridden and sometimes hungry." And he's quite right. With the blood of my daughter virgin no longer on his couch.

I can see that whole thing. A girl I know whose hair was so golden that she brought actually the outside sunlight into my alchemist's den when she brought in my lunch. She was laughing all the time. And I felt more confident than I've ever felt in this life. I was just on the verge of intellectual concepts. What I was doing with mercury and lead was only a fulcrum. But what made me confident was that I was skillful, I could swim and dance and use tools, make my own. And that girl laughing wonderfully. That was a long time ago. In a way it's a dream. It's as strong as the time Buxtehude came in here, an irascid old man, two hundred years and more out of his time, confused, irritable, fusty old clothes, everybody laughing at him, in all this stretch of mechanized canneries and impossible times, out of another language, and a German not even spoken anymore recognizing only one thing. A phonogr. record of his chorale fantasy "how beautifully shines the morning star." Fabulous organ, he'd never heard such a thing, must be in the loud speaker. Stamping up and down. What a crazy business. I must have told you. Had me finally in a real sweat. I *had* to get that man to the Princeton University old-organ-replica. Right away. By plane. And I had to find not only some really proficient German scholar, but a scholar of 17th century German. I got to thinking of Blinks at HMS and then Van Niel. What a relief to run onto Van Niel. Just from toying with the idea. Hell you don't toy with an idea like that; it gets *you*.

Like where I got by the idea of nostalgia. The world's most powerful motivation. It relates the thing not to its use, but to the powerful emotion. I guess of security, anyway of beauty, that gets associated with it. No one knows that. Well you know it. Thomas Wolfe certainly did. Power and profit and love and hate are the motivating forces. Who realizes the power of nostalgia. You can examine art, business as a system of power, government—especially government, religion. Science a hundred years from now. Right now it's too vigorous, no one's powerful enough to hold it now to nostalgic dreaming. Or bureaucracy best of all. In any institution, socialist or profit making, no difference. At the base there's that great pyramid of the way things used to be, in the good days, the good old days. "These are the days of our youth, the days of our dominion. All the rest is a dream of death and a doubtful thing." And it's really true, that's the funny part of it. And I'm tied up to some curious vision of golden hair, though I don't like blondes particularly. Well what strong thing are you tied to mister j.sbeck, mister brook mister stone brook which I haven't seen for a long time?

Best thing I can do now is to go back to designing what can't be done; a printed form for the Charlotte Ids. survey, a form accurately adapted to convenient use with the typewriter. You better wish me well because I sure as hell need it.

To John Steinbeck
August 1946

Couple of things occur to me as interesting-amusing or both. The truck that people gather around them I suppose is their life in epitome. I was thinking about the confirmation and baptism certificates that mother carted around with her from Massachusetts to Wisconsin to Chicago to the Pacific Coast; the funny old greeting cards of 75 years ago. Mostly I got a fine nostalgia out of Frances' report that at the bottom of mother's trunk there was the funny old fashioned voluminous dress, entirely handmade and so ruffled and hemstitched, in which I was baptized and I guess Frances and Thayer too. How terribly unfortunate that the generations are so discontinuous. That my children and Nancy Jane's (she's going to have a baby this month or next) couldn't be baptized in the same one. Like a Pope's mitre; how charged it becomes in time.

And another. Directly after mother was taken very sick, she wanted an Episcopal priest. Terribly unfortunate that a few months before this, the satyr in Rev Hulsewe caught up with him again. He got to fooling around with little boys. The wicked old women, of the church that was founded by charitable Christ, turned on him viciously. As of course they would. The mothers would, understandably. But worse the old maids who had nothing to lose, but who were afraid of sex even in its normal expression. So out he went, an old and broken man without—now I suppose—even any money, and most of his Holland relatives dead or scattered. So that real friend of hers, as she still of his, wasn't there. Frances suggested Rev Clay whom we all love and like. Mother was maybe a little out of her head then already. Or maybe more in it than she'd been ever before. So she said "But he has such a long face, I'd like someone more jolly. Can't we have someone who likes to kid?" Already when they got someone she was out, but I suppose it did her as much or as little good one way as another. Consciousness isn't everything.

I really didn't have a bad time at all. I have enough training as a nurse—tho not complete and not very good—to have the MD's regard for a very sick person as someone to be worked on, like as specimen. Not at all lacking in love or consideration or even personal feeling, but just that notion of scientific doing-well-for. A deeply unconscious person is like an anaesthetized specimen. The things we know and like are mostly in abeyance.

Well, I never liked my parents too well. Now they're both gone I feel pretty much the same, but with regrets because I didn't. Not personal regrets. It's the pace we're moving at. The generations aren't connected and they can't be. But it's a pity. Both lose so much.

Frances sent over to me among other things an old writing desk that mother had since she was a child. Gosh how I lusted after it when I was 5 or 6; probably because she had so much affect charged into it. They certainly made things to last in those days. But now it's only a vehicle for nostalgia. And to make me wonder what happened to all the potential of that undoubtedly potential woman. Probably beat out by reality on the middle level; now all tied up in that writing desk as mine is tied up in the ghosts of letters to Irene and Jean. That Ed is certainly a jazz specialist. Armstrong Mahogany Hall Stomp, Mezz Mezzrow, Jellyroll Morton, the Missourians, a fine bunch taken over long ago by Cab Calloway when he was good, wonderful Duke Ellington still.

And a Chinese Idol I still like. But most of the folks' things were gone, frittered away, broken, given away, thrown out by Frances when her different and then immature tastes took charge. Before they moved to the more fashionable apartment in Chgo, Frances had the junkman cart off—now in memory I bet she kicks herself—some commodes, tables, etc, of carved chestnut, marble tops, elaborately scrolled mirrors. Worth a lot now as antiques. I had the beautiful chest, I guess a highboy, that grandmother used as a curio cabinet. And I did too. Grandfather's sea chest; why I don't know; he never went to sea. I guess in those days it was a good thing to have a sea chest. All burnt. With so many things of my own, and with daguerreotypes of both families spelling help. My mother was certainly good looking and feminine as a girl. Not like Janice's, but plenty OK still. I wonder what possibly could have happened to all those warm characters. Got beat out by her executiveness I suppose. By unrelaxed marriage. Probably all of us children were welcome, but I went my own solitary way not sensing it particularly and now I feel a little cutoff. Like Goethe in the *Faust* dedication all alone. Well of course that's it, isn't it, what you wrote about. That feeling of aloneness. Those feminine imprints are certainly put onto a person mighty young; no wonder I like gals with big eyes and thick lips. Everyone of course has to have a mother image to love. And if you don't particularly like your mother, poor woman what a fate for a woman, why then you're particularly lost. But I still wonder what happened to all that gorgeous potential. And because I think I can even work out in my own head and correctly what's wrong with the world, tho how to cure it if I were God himself I wouldn't know, makes a person perhaps feel lonely too. What's wrong with the world is nostalgia conditioning that fears change, nationalism and its handmaid xenophobia and its father power-drive, the profit system and the propaganda and advertising of special interests that keep it in the saddle safe from change, the mechanicalness, the engineeringness that rides us, plus the usual personal greed of black markets. Science is pure, but *unbalanced* by being engineer-driven. Unbalanced anyway. Religion is deadish or cultish or wishful. If science leads us where are we? And science leads us alright. That editorial of De Voto in the July 1946 Harper's Easy Chair is very very good. Maybe the best thing is to close the plant and go fishing. But I think perhaps that's been done too much by too many good people while the others build an empire on carelessness and uncitizenship and prejudice blown into intolerance. But who says what's wrong with the world is the avoided bringer of bad news. Bernard Shaw that ass to the contrary notwithstanding. Still, he's an ass about only a few things. How he rides the antivaccination, anti-vivisection hobby horse. But also how he does look on the avoided minority side of almost every other question. Incidentally even you couldn't have done better with prejudice. First thing I know I'll be looking it up.

It's funny that your book will get into me though I don't know it; my nostalgias second hand into you. Norman Geschwind here has some good ideas, tho so mixed up as a person, such a curious sly arguer. Still he has a slight "ultimate reality" back in those ideas somewhere. And Jorgen Bering, for all his fascisms. So hard to talk to, so easy to fight with. People that give their life up to an idea that they're sure of. Is fine. And what's wrong with the world is of course what's wrong in epitome with every individual in it to some degree, mister 'a' almost completely avoiding the pitfall which almost completely capture 'b,' and Lao Tse almost avoiding nearly every one of them. Ed Jr just came in here with an almost perfect negative of an isotherm chart that Hedgpeth drew for the (when?)

new ed of BPT. I said "It's perfect" Ed said "Isn't it! But I'll get a better one!" wotta guy.

I suppose in re-reading, parts of it will not hold up well. But as the adventure of one man's merging with man, it struck me forcibly. The guy knows about what I've thought of as breaking thru. In fact the book is an exposition of that emergent. And about participation, he calls it love. The things he's interested in are, literally, the things that have no name. Which makes me for the moment feel good about my own externally useless place in this war movement. Which is actually another one of those migrations of the human spirit. With which all those go along who have any contact with the human spirit. Which permits such a person to have as good a position inwardly at least as the soldier or director or munitions worker—in even a larger scheme, in the structure that underlies both peace and war. Still, I might feel even better if I'd get in some CD [civilian defense] work, which I must do next week—said he putting it off again—even tho Norm says it's a mess, and better that good people should wait until the next (more significant) thing comes along.

Well, say hello to Gwen, and write as you get a chance

⌒

To Betty Farmer
August 19, 1946

Dear Betty:
Nothing was the word because we have been so confused. That is, Toni is confused. I'm never, of course.

And for several reasons which I will detail. One important. Kay had another attack of her trouble, the Dr. in the meanwhile insisting that his diagnosis was in the main sound—and X-rays to bear him out—but not malignant. (Proven by the fact [that] she was still alive). So this time the sight of one eye almost dropped off, and very rapidly. Since the risk of death in operation was better than the certainty of blindness, and probable eventual death without operation, we went ahead. Last Thursday. I went up there. Not that I could do any good. I just sat out in the hall. But it might have been comforting for Toni. At least that's what I intended.

When they got in the skull, they decided that a two stage operation was necessary. She rallied from it nicely, with some unpleasant breathing sequelae due to irritation from 5 hours breathing ether. But that was eliminated first by [an] oxygen tent, then by making an incision in her throat for putting in a breathing pipe. The next will be in about a week, when the tumor will be removed. She has a good chance, and if she makes it, she'll probably [be] quite OK. In any case, everything is being done that modern science can do, and love, and every care. Fender, [professor] of neurology at Stanford U is in charge; he likes Kay personally, he likes all of us, and all of his considerate efforts are being brought to bear. The other night when another specialist thought the breathing thing might be due to brain pressure, rather than local irritation, Fender, that good man, came over at midnight and slept in her room until 7:30 so he could be right on hand to check symptoms.

The operation itself was done so skillfully that Kay, who fears the word, doesn't even know she was operated on. Doesn't even remember the anaesthetic. Thinks she had a particularly long X-ray session which made her head irritated and required bandaging. Woke up promptly after transfusion and demanded a funny book after saying now she wouldn't worry about X-rays anymore, that that one didn't hurt her.

So. You see. Many of the things that T intended to do, spoken and unmentioned, haven't been because of this. She had written more for *What's Doing* etc, some of it published, and will undoubtedly send it to you. Hasn't started on writing about the northern trip yet. But will do so soon surely in any case. Resting up first (because no nurses have been available, and she's been substituting).

I got the materials which I collected up there, whipped into good shape quickly this time, and reports are beginning to come in. A number of new species already. I anticipate that from 1945 and this summer, I will have turned out 20 or 30 new species. Altho that wasn't what I was after. I was simply trying to find out what occurs there. I haven't yet settled the matter of the two species of horse clam; very curious. Sometime I will know and then I'll write you.

We had [a] wonderful time when first we came back. Our house was full of happy happy people. Not too sober. Cooking wonderful meals, welcoming us back. I found the chemical lab work very pressing and very piled up; fellow had to come down from Research Dept. in the city in the meantime, [in] addition to the guy we had pinch hitting. So I have been busy on that score too.

Well, so I'll let you know on the Kay prognosis. Sardine season opens tomorrow, price adjustments finally made between the CPA, the fishermen and the canneries. I anticipate a busy season. So I am writing this while I can and will post it now. Tell Bill to keep the squid thing in mind. And keep planning on coming down here because everything is very pretty and will welcome you. Even fog. Even beer.

Ed

Betty Farmer owned and operated a hotel in Clayoquot on Vancouver Island. She and her brother, Bill White, befriended Ricketts and Toni on their 1945 trip to the Outer Shores.

To Nancy Jane Ricketts
September 13, 1946

Dear Nancy Jane:
It was all very good news about you being OK, and about the child being a boy—what did you name him?—if it was a boy that was wanted and I suppose so. And that now you are so nicely established in a home of your own.

Apparently there is no money problem in your family, luckily. I suspect among many other things you have financial security. But perhaps you can use the enclosed ten dollars for something not budgeted, something you wouldn't have otherwise; a hair do if nothing else; or for the baby, a luxury if you'd rather. Tho you don't talk about luxuries nowadays in terms of [a] mere ten dollars.

[In] spite of working fulltime again this fall, I've been taking it fairly easy. Not much fish, not much work. Kay has been sick again as perhaps you heard. Has passed successfully the second of a dangerous two stage brain operation and I think next week she'll be out of the hospital—Stanford Lane in SF, and Toni will be back down here again all the time instead of just weekends. Cornelia was here Wed. evening; we had a very lovely dinner at Angelo's, and music the rest of the time. She's certainly getting along in her music, both singing (has a chance at $50 per week for a couple of weeks in a SF theater) and appreciation. Ed has gone back to school. San Francisco State College. During mother's last illness, I was over at Fred Frances a good deal helping them look after her—they couldn't get a nurse, and I saw the lovely transparencies they took of you in your going away clothes. Also Frances and I got to be better friends again for that time. For a while she hadn't seemed to like me at all, never did know what it was all about, something about Cannery Row in the first place apparently. But since it was something in her near as I could see, I didn't be concerned about it, supposing eventually it would be worked out. Or not. In neither case not much I could do. And it seems fine now. She and Mother got to be quite friendly in the last few months, and on the whole mother had quiet a pleasant blooming in her old age. And to climax everything else during this difficult time, both my car and Fred's went on the fritz. What a time! Do you remember Joe Campbell? He and his wife were here overnight. We had a lovely time. Toni came down from looking after Kay for that evening; went back on the morning train, but will be down again tomorrow night. Tal is about to have another baby any minute. Her third. She certainly likes 'em. I'm having fun with working up ideas, with playing with the animals I got this year and last up north, and with the records of them, getting out a few orders. About enough to keep up [with] taxes, or so I tell myself. The Bach Society has been made a permanent all-the-year institution here on the Peninsula, and the chorus will practice and sing the year round now. I think it will be a very lovely and very famous organization eventually. Like the Bethlehem chorus. Well, have fun Nancy; don't work too hard. To contribute one's share of fun and happiness to the world is certainly important. Actually a real duty, especially during these hard times. Hard I mean in the sense of war-like and uncertain. I think never in the history of the world have the forces of fear and prejudice and intolerance threatened our civilization so terribly. And no time has it been so necessary for people to think and to love so wisely; and at the same time there's never been another age when there was so little of these. I am going to apply for a Guggenheim fellowship and this time I think I have a good chance of getting it. Then I can continue on the summer biological surveys up north. Queen Charlottes again next year I hope, and then onto the Gulf of Alaska. Eventually the Aleutians. Jon is writing this time apparently a very worth while book. I hear from him often. Bob Rossen was here last night too. He and Toni and Lewis Milestone worked on a movie here a few years ago, and now he's become a director in his own right. Well, I started to close this letter a long time back. Very tired; things didn't break up last night until nearly dawn. My good love to all of you; the new little grandson, Perry, you.

To John Steinbeck
September 15, 1946

Dear Jn: I'm sending along a copy of *Circle 9* with Toni's article on Ellwood (signed A Seixas—what psychopathology prevents her signing her presently known name is beyond me, but there it is). She develops some ideas on art and criticism that are new to me and I think quite good. I think it is, incidentally the most meaningful writing in the magazine, but then of course the whole thing isn't much concerned with meaning, but mostly with visualness or hearingness—of words. Tho the first group of poems, Melissa, I started being delighted in by the flow the words, and then discovered afterwards that there was meaning there too, altho I didn't stop to work it out.

Also if I can pick one up, the Sept. "What's Doing" with Toni's article on Peter [Ferrante], I'll send that along. The magazine situation here is such that if you don't grab an item when it first appears, you never do get it.

Kay comes out of the hospital tomorrow, and later in the week Toni will be here uninterruptedly. She came down for the week end, and earlier once when Joe and Jean Campbell were here. They're fine. We had a lot of fun. Jean is a beautiful no I should say a significant dancer. Even I who know nothing about dancing could recognize the . . . whatever it is, what in photos or documentary films I would call verisimilitude. The same evening [Robert] Ross Rossen showed up, the film writer who works with Milestone. I guess just lonely, he went up to SF the following day with Toni, then flew back to Hollywood where he just finished making a picture. A director now I think.

I am interested now more than ever in comparing the action of human society as is—and how it got there—with the presence of societies in the tidepools, and their controlling environmental factors, but I don't know if I can work it out intelligently, acceptably, and interestingly. Ed Jr is installed at SF State College, seems to be handling everything so far very maturely. He can't get in CIT [California Institute of Technology] anyway until next year; the competition even then will be huge. The waiting list is so great that they have the choice of a lot of mighty fine minds, and of course they pick what according to their standards is the most acceptable. The approval of CIT is the only thing I can help him with.

Ed

The article Toni wrote was "Civic Citation—Peter Ferrante," which appeared in the September 1946 issue of *What's Doing*.

Robert ("Ross") Rossen was a New York playwright who moved to Hollywood in the 1930s to write film scripts, including *The Roaring Twenties* (1939), *The Sea Wolf* (1941), and *The Strange Love of Martha Ivers* (1946). He made his directorial debut with *Body and Soul*, which was released in 1947.

Ricketts's frustration about the overfishing of Monterey Bay is apparent in the following letter. As the sardine population continued to decline throughout the 1940s, local concern grew, and the *Monterey Peninsula Herald* began publishing an annual "Sardine Supplement." Ricketts wrote the primary articles for both the 1947 and 1948 reports, commenting on the various factors impacting the population and causing its decline.

To Ritch Lovejoy
October 22, 1946

Dear Ritchie:
As I suggested today, I will try to work out a statement below of what I think the sardine situation amounts to. But it'll be long and rambling, and thinking out loud, hell I don't know if I can even organize the thing enough to make it make sense.

The sardine is a migratory animal. As an adult it can eat only plankton, and usually the smallest plankton, chiefly diatoms. It stores food in the form of fat, and all that fat is converted into sexual products. Its range is limited to a path some 1500 miles long and never more than 50 miles from the shore, except in breeding time when it heads out to sea off southern California into the warm water. It, in common with all pelagic life, certainly has its annual fluctuations even normally; and they'll be large ones. Certainly its only food, plankton, has huge fluctuation in mean annual productivity. Its breeding place is varied somewhat by the temperature of the water, that is, it goes into water of a certain temperature to breed—rather than to a certain place.

Distribution of animals is forced probably chiefly by population pressures. Migrations are not without cause. You'll notice that as long as things stay prosperous in the old homestead, people stay pretty well put. But when the farm is divided into smaller and smaller units, somebody has to move. Potato famines in Ireland are not unconnected with pulses of immigration to this country.

If it weren't for competition for the limited available food within the area of their origin, I'll bet the sardines wouldn't ever move. Except maybe for mild peripheral driftings. But here comes along a particularly heavy crop of eggs and young fish, after some particularly favorable plankton years. Most of them get eaten up, but enough continue to grow to make the food supply seem pretty skimpy. The young ones aren't strong enough to swim far and fast. They have to stay there and take their chances of starving. But the older year classes, stronger and able to undertake longer migrations, get to moving further and further out. Some of them get to going up north where they find a particularly favorable set of conditions.

In the north, most of the plankton production is in the summer and fall, but in Sept., and October particularly, there's a peak up north (say from Oregon on up to the Gulf of Alaska or to Queen Charlotte Sound) that we don't have at all in the south, southern California especially. From the 37th to the 44th week, almost 25% of the whole year's diatoms are produced, whereas in the La Jolla region only 5% are produced in that time. So if they have to go anywhere to get food, it [is] obviously pretty much to their advantage to go up there.

But now suppose a year or two go by with lowered plankton production. The old ones now particularly have to go north, but none of them make out well. The young ones don't grow up into sexual maturity very fast. The old ones get very little excess fat that they can devote to sexual products. So the next year's spawning is a flop. And maybe the next one too.

Now then in the following year there isn't a very heavy population pressure to force any migration. Especially there aren't any very young ones that have to stay put, and that therefore force the big ones north. There isn't much of any migration, all of them mill about where they are, especially if the feed is good. And by that time the plankton probably has started to build up again, and there's still that further reason for not migrating. And such of them as do go north, don't get up there so far, and drift back earlier.

All this would represent a pretty normal fluctuation. The animals would in the long run hold their own, because as soon as the situation eased up a little bit, those of the old ones that hadn't died (and they live a long time) would be able to put on lots of fat which again could be turned into huge numbers of eggs and sperm. And in the meantime the few small crops of young ones that had been born,—those of them that survived their enemies and the bad times—would just start booming along, achieving sexual maturity in record time, piling up great chunks of fat. All of which would convert to sexual products. And there'd be a bumper crop. And then if the little fish found good conditions for nourishment and if their enemies weren't terribly more abundant than usual, in a few years you'd have a crop of sardine to gut the canneries.

But in the meantime it isn't working that way. In the good years, tons and tons of fish are being reduced, the canneries get larger, there are more of them, they spread out to the west coast of Vancouver Island, to the Oregon coast, to Ensenada, the fishing boats get larger and finer, there are more of them, they're better equipped, and the fishermen themselves become more skillful. So finally an industry gets built up that not only can handle these bumper crops, but that has to have them in order to operate.

So now there's another factor working on this up and down graph of normal sardine production. And that factor pulls down every high a little, tho not so much as it pulls down the lows. And when you get a time that would be fairly bad anyway in the life of the sardine flocks, it makes it disastrous. It could conceivably sometime hit them so bad, when they were at a trough anyhow, that the margin by which they survive over their normal enemies and hazards would be wiped out completely, and they'd go down. If not to extinction, at least to commercial depletion.

That's what I think has happened to date. Of course in the long run it won't do much harm in any case. And this time I think they'll come back, starting a little next year. But even if there *is* commercial depletion, it'll be only temporary, two years, five, ten, fifteen at the most; good thing for the sardines, they'd probably build back up to strength again good as ever. But it needn't be; there needn't have been anything worse than a trough in the chart for a couple of years if we'd got to know something about the beasts, planned things out a little. The fur seal was brought back into greater production than even in the palmiest days of international poaching; and it was on the way to complete extinction. The halibut is coming back into steady annual or predicted fluctuating production. The salmon would have been saved in the US if it weren't for the lobbies of mining interests, and for unrestricted dams; and they *are* being saved in British Columbia and Alaska. Sweden is producing now more lumber annually than they produced a hundred years ago; compare that with our forest wastelands! But in the long run, the sardine perhaps won't be hurt. Our policy of insisting on taking everything at the moment, free enterprise at its freest, hurts only us. A few fishermen and their families will suffer; they'll lose their boats; some of the canners and reduction plants, the new ones especially, may go broke. The town will be hit pretty hard. It'll all bounce up again. Even if there's commercial depletion for many years, as there seems to have been in France, most of the people involved will survive somehow. They all bounce up again. The sardine too.

But it needn't have happened at all in the first place with decent planning. Or at least it needn't have happened so badly. The industry shouldn't have been allowed to build up so fantastically. Here, in San Pedro, in the Bay region, even in San Diego and Ensenada—and along the west coast of Mexico where tuna fishermen take such hordes of young ones for bait—on the Oregon coast, clear up the west coast of Vancouver Island. The number of plants have been doubled in the past few years. And the sardine hasn't doubled!

But it *has* happened. And it'll happen again. Because there won't be any decent planning even now. Already the operators are protesting the commission's curtailments. "It'll work a hardship on us" they say "if you cut down our permits for open reduction." Why can't the commissioners say "It isn't we who work hardships on you, it's the lowly sardine. And they'll do worse, not we. They'll wipe you out in fact if you don't apply a little common sense in conserving them. A very little less greed now may mean so much more profits in the long run!" But nobody'll listen! The industry says "This can't happen to us. We refuse to believe it!"

Ritchie. PS: The sardines they're getting in San Pedro now, that are so big and fat and full of oil, are of course the ones that migrated past here in August, earlier than usual, because they didn't go so far north. (Because there weren't so many of them as usual, due to the many causes I've mentioned, including overfishing. And therefore there weren't the heavy population pressures to force the usual migrations clear up north). That's actually the crux of the whole thing right there.

⌇

To Joseph Campbell
October 25, 1946

Dear Joe: Many many thanks to you for all the fine things. The pretty things. That legend of the Jew and the soldier at the bridge is certainly something. Seems like you're blossoming out as a writer at a great rate. I have to readjust my mind to think of you that way. As a thinker and a teacher, but not until now as a writer.

I have been looking thru the key again in connection with picking up the *Wake* from time to time. I certainly agree with you that those last few pages are among the best, quite *the* best, in modern literature. Perhaps in any western literature. How such depths can be evoked thru the printed word is more than I can see.

Things happened since you were here. Xenia of course came on. We haven't seen as much of her as we hoped, but she's kept busy looking after the kids, and we may see, oh of course we will, more of her before she goes on. Changed. More level. Still thank heavens you can't even prevent a screwball from being at least a little screwballish. And Toni came back with Kay from the hospital. Who is doing fine; seems to be suffering from some anxiety and insecurity, but usually handles it well. Of course she's really been through a lot.

And my work on the little animals goes on. I applied for a Guggenheim. Think I have a good chance for it. The Foundation people themselves finally suggested I apply, after Schmidt of the Field Museum had apparently written them about me. So I may have another good trip next summer despite hospital bills.

Very little Cal Pack work here this fall. And of course I put in the added time to good advantage otherwise. Everyone very downcast about the sardine situation. And well they should be! Just another instance of greed, lack of foresight, lack of seeing beyond a narrow individual segment of a large picture. For years the canners and reduction plant operators and fishermen have been warned they were taking too many fish. They refused to listen, selected their evidence, petitioned for more and more permits, put pressure on the Fish and Game Comm., lobbied the legislators, always got their way. And now so sad. No fish. Of course they'll come back, the situation will straighten itself out after a bad time. But it's all so unnecessary. Need for the fish has increased many fold during the past ten years. But the fish themselves haven't increased in any such fashion.

Like the auto traffic situation in NY, per *Fortune*'s recent survey, it's all tied up with the refusal of the average person to see any picture larger than his own. Then when the social pressure hits him, he wants to put the blame outside, never take his share. I think of that very often in driving. I come into an intersection, look in the rear view mirror—unless I'm terribly in a hurry—and consider how many cars on the other street can go by unstopped if I pause a second longer. In other words, just what the crossing cop has to do. If everyone would do just the least bit, a greater good would be served for a greater number, including the guy himself on some other occasions if not then. But damn few people ever consider that. And if they won't do it in a thing like that, how much less so in larger and more important items. So the lumberman takes all the timber and beats his children out of an equal share in forest products of the region. And all the while the Swedes, thru conservation, get now each year more out of their forests than they got 110 years ago when the present program was instituted. Now the same thing happens to the fish. And it makes it so much harder to explain, when you know that there are peaks and troughs anyway, and that even if they didn't take too many there'd be some bad years. But not so bad, of course. And not so disastrously long. Here this year, many fishermen will suffer, their fine boats will be foreclosed, some of the new reduction plants will go broke. A pity they were allowed ever to be built, especially at this time with difficult materials priorities. And people go without homes of their own while the unused reduction plants promise to stay idle. Now I see one of the social values of organized religion. I've thought so little of them I never realized that before. If people won't of their own accord achieve a sense of balance with something suprapersonal, you beat it into them. Well, I'd rather be an ecologist and have for stock in trade that sense of integration with a whole picture without which any field zoologist must be lost. Or, I suppose, an integrator of myths and literature.

I think Jean is fine. If she wants a vacation from concert dancing and from New York, bring her here for awhile. She will go well with this bunch, like and be liked. If you people would spend a summer here, I think there would be many good times, and you'd get some work done. That fine Wings Over Jordan negro choir comes here tonight; if they still have the wicked soprano it'll be good. Well, why don't I get to work. Read a vitamin or two. Assay a protein for my living.

Ed

To Pascal Covici
October 25, 1946

Dear Pat:
Many thanks for the Perkins Roosevelt book [I] just received. And for the Viking Portable before that, which Toni is still busying herself with. I think the idea of the portables is a very good one, combines the advantages of the one volume Blake, Donne etc which were so valuable with smaller size and greater successibility. Apparently things go on reasonably well there, despite shortages of paper

stock etc which must be very frustrating. I finally went out and got myself a *Finnegans Wake* (after looking again thru the "Key" that Joe Campbell sent last year). A fine thing those books are going well. A good thing that anything so valuable and permanent as Joyce should be salable too. Almost popular. Reminds me of the situation with Goethe, particularly *Faust*.

Jn was in a fine fettle these last few months apparently. I haven't heard from him so frequently, and pleasantly, and most of all, livingly, for a long time now. I gather the new book will be really something. Probably good from every standpoint, including profits—which after all aren't to be sniffed at. Well, he writes that you're in good condition too. So the bad world times at least can't get all of us down! I go ahead working at a great rate. Applied for the Guggenheim ten days back. Best of regards from Toni and from me.

Ed

Steinbeck's *The Wayward Bus* was completed in October 1946 and published in 1947, receiving mixed reviews. Edward Carberry wrote in the *Cincinnati Post*, "*The Wayward Bus* is not intended for the young or prudish. But the coarseness is, like the people, cold and separate and unexciting. It is as though Mr. Steinbeck were a biologist and his characters so many specimens" (February 15, 1947).

To The Gramophone Shop
November 2, 1946

Gramophone Shop Inc.,
18 East 48th St.,
New York 17, NY

Gentlemen:
I have your card of the 30th Oct with the news that Vol. 1 of the *Anthologie Sonore* is out of stock. When it comes in, will you send it on; or let me know?

And with the sad news that all the others are available. Sad because I must have now more than the one, and who am I to be able to afford such luxuries!

So another check for $23.95 is being included. Will you send me

Vol. 4 and Vol. 5 of the Anthologie	$40.00
1942 Ed. Encycl. Recorded Music	3.95
	43.95
less check attached previously	23.95
	20.00

Sincerely,
E. F. Ricketts

To Theodore Seixas Solomons
January 22, 1947

Dear Mr. Solomons—I suppose Teddy is better—Kay's pet name: Toni set for me the pleasant task of writing to you. That way it'll get done. If we wait for her who has little leisure, maybe it won't. And I am away from the house too, and free on that score, which she isn't now.

Kay has had a couple more little attacks, and I suppose we have to face the fact that pressure is building up again and that her troubles definitely aren't over. A dear little wormie who's so cheerful and affectionate. Learning the discipline of handling pain and franticness. Our problem is to prevent both of these for her much as we can. No immediate crisis, and I think there won't be one for some considerable time. She just gets overexcited now, can't hold herself down, and we have to do that with phenobarb etc. We had a bad time couple of nights back. Toni didn't get much sleep. Last night was better. I think Kay dips in and out of these troubles, that incidental things like too much activity, or getting overexcited help to bring them on, and these we can reduce to some extent.

She suggested originally that I call you, then thought writing would do as well. People who like each other undoubtedly can help each other at difficult times, and without themselves taking on any load—with actually the reverse; with illumination sometimes. Though I suppose it puts out energy. Anyway I suppose that was chiefly what she had in mind asking me to write you. And a sense of contact,—which that little bum maintains not too well since she writes so seldom.

I have a grandchild; my middle child Nancy Jane had a boy, Perry III, few months back in Baltimore. Ed Jr. is in and out; he's going to school in SF: chemistry, math, physics. And my youngest child, Cornelia, spent New Years with us. I am up for a Guggenheim fellowship, with good chances. John has written a new book of perhaps some significance "The Wayward Bus" which we read in galley proof. The lowly sardine has been missing from his usual central Calif. haunts. In the San Pedro area there has been greater number than usual. But still not enough to make up average total landings, and it looks bad for conservation in California. And sad for Monterey which is above all a sardine town. So I have been not too busy here in Cal Pack lab. Had time to do lots of my own work. The northern survey, and my study of the sardine. Except for this recent upset and some hypertension, Kay has been fine. Happy and active. And we do well. Toni—Eleanor the little philosopher takes things very well. If there are any developments we'll let you know.

Ed

To Joseph Campbell
April 1947

Dear Joe:
Very good news. I'll certainly be glad to see you.

We'll probably be away when you come through June 30th. But back long before Sept. 11th.

Expect to leave here about May 22nd for long collecting and recon. [reconnaissance] trip—something like we took that good year but not so elaborate. West coast of Vanc. Id. (where we spent last summer also), Queen Charlotte Ids., Prince Rupert, perhaps on up to Stewart-BC-Ryder Alaska. Then I think I'll be finished with the Canadian outer coast line. Gulf of Alaska next. Then when, if ever, I get out to the Aleutians, Bering Sea, maybe Kamchatka; I'll have covered the entire area from Panama to Asia well enough at least to start on the manual. Almost 50 now; give me anyway twenty more years of good health, not too much restriction financially and I think I can do it. Or lay a good foundation for the next guy.

Yes, I've been in the army, working since at a regular bread and butter job. More or less. At regular hours. Biochemist. Which should amuse anyone who knows how I hate math, and how little chem. I know. Now on part time, and building up PBL again as a commercial supply house. But back again for a season after my summer leave of absence, and when the canneries open up full blast. I have been on biol-fisheries research also, mostly for myself, partly for Cal Pack (cannery) and I've certainly dug up some amazing things about the plankton. By what magic the annual peak of a 7 year mean curve for S. Calif. comes at precisely the same week as that of a 6 year curve of different years in the Aleutians is more than I can figure. If we had long period data on this, all sorts of ideas might be justified. Just now it all seems pretty close to the primitive jumping off place biologically; anything could happen. A marvelous calendar for the little bugs to coordinate into! Trouble so far in understanding it has been that the single year and the single locality variations have been so immense, no one bothered to construct a mean curve.

Ed is back from three years in the army, most of it overseas. He's fine. Wonderful photographer, wonderful musician, wonderful mathematician. Going up North with us too, and then into Cal Tech on the GI bill of rights if his qualifications fit. Did the most interesting army work I've almost ever heard about: one of 5 [?] and 4 officers who had charge of the signal corps annex of the SWPA invasion plans. Got a T/3, worked his own hours, lived mostly in hotel, worked sometimes very hard but of course with the greatest sense of significance and drank apparently plenty of beer. Toni has come here with me since you went thru last. As Xen no doubt will have told you. A pretty cerebrate gal; I imagine you two will get on fine, but those things are beyond prediction. Anyway come on, we'll be so glad.

To Joseph Campbell
April 11, 1947

Dear Joe:
Your letter lifted Toni more perhaps even than it did me. She was a very tickled gal. "To be included, so nice" was her note.

Sign on PG church announcing sermon: "After Easter, what." Did you like the article in recent *Life* on Medieval Europe. I was surprised how clearly they got over what I think is the substance of those times. Very significant that we, who among nations move most swiftly to the break-up, should have ourselves destroyed Mt. Cassino monastery, Thomas Aquinas' nurturing place. We destroyed our grandfather's house. I fear there'll be many a bad time before our children build its equivalent.

A thing you'll like in the new Ed. [of] *BPT*. The plankton essay. Or did I tell you. If not, you'll be interested. If so, only a little bored.

The man who for thirty years has been laboriously counting the organisms per typical liter of sea water daily, and then composited weekly, finally retired discouraged. The fluctuations (he himself calls them pulses) vary from a few hundred to hundreds of thousands within a few days or within a few miles. The university refused to publish any more of his figures after seven year reports. But did publish some of the summaries—which lacked the details some of us would have liked for other combinings. Anyway, after investigating the limitations of mean curves for this material, I constructed [a] 3-station-composite means curve for the 7 year period in S. Calif. Out of data from a single station in the Aleutians for 6 different years I also constructed a mean year's calendar based also on daily readings. These both came to peak in the *same week*. There was a huge secondary peak in the north (reflecting the usual northern autumnal re-growth which doesn't occur in the south) but there was a corresponding but smaller polygon peaking at the same week. The 5 or 6 smaller humps all came to peak in or near the same weeks! Isn't that fabulous. If substantiated by the 25 year figures, which have been got and could be made available if a person had the money to get them tabulated and worked up, this would reflect a primitive biological rhythm operative over the whole of the north Pacific, the same presumably in any large group of years.

So we both strike fertile ground. For you're doing so certainly. And sounds as tho you're at it steadily. I *did not* get the Guggenheim. A pity. I was counting on it rather, since they suggested I apply. That'll make this summer different. I had figured on going again to the Queen Charlottes, then taking that three year's work as basis for a book "The Outer Shores." Now it'll have to be deferred a year, perhaps indefinitely. Kay's operations and a couple of other unexpected items see to that! Toni did a job on Jeffers, also another little thing called something about are we losing our minds. I'll send them on. Now she prances around here half clothed, well not even half, saying come on now get ready to go to Barbara and Ellwood's. I forget if you knew them. They do well. He traded paintings for a Packard Station wagon. And has now an entire block of wooded hill near his house and some added lots. I have just bought two of them and will hope eventually to put a lab up there, since I can't get waterfrontage for love or money. And this is too valuable and high taxed for me to hold onto indefinitely. Oh that impatient gal. Here we go. To Jean and Joe from Toni and Ed.

In the April 1947 issue of *What's Doing,* Toni's article about poet Robinson Jeffers entitled "The Hawk and the Rock" was published along with "Are We Losing Our Minds?" in which she analyzes the way fear is used in marketing and advertising.

To Nancy Jane Ricketts
Summer 1947

Dear Nancy:
If the impulse to send me beer becomes too overwhelming, it can be done by express, perhaps however even a little more expensive than the parcel post 12c per pound rate, so not to be entered into too lightly.

My 50th birthday was really fabulous. Hard to surprise me, but I was then, until almost the last minute. The living room was decorated with a great scroll. A draughtsman friend put in lots of time on it. Alleged letters from the WCTU, the American Legion and various reactionary outfits all congratulating me. A card from the United Brotherhood of Cats. Very fun. More than 20 people packed into that little room when we opened the door.

The doll I had forgot about. No sooner said than done. Toni just got it out and put it on the floor next to the front door and I think eventually you'll have it.

I finished up the [2nd] Ed. of *BPT* which was a big big job. Eventually I suppose it'll appear. I planned another book to be called "The Outer Shores" to be based on the last two and the forthcoming summer's trips to the Queen Charlottes and the outer shores of Vanc. Id., but that'll have to be delayed at least another year because this summer we can't go north as projected. Kay's hospital etc bills took too much. She's fine incidentally. Jon was here, and before that, Gwen for a couple of days. We had a fine time. He seemed to be his old self. Since then he had an accident, soon expect to be out of the hospital and on crutches. The children are wonderful.

Get such volumes as you can afford of the *Anthologie Sonore*. Gramophone Shop, New York. Twenty dollars for volume of 10 records.

Fine mostly pre-Bach music. Really very fine. Old madrigals, masses, troubadour songs etc. One was given to me for my birthday by the gang. My phonogr. still sounds wonderful, but I heard a new loudspeaker recently that backs mine off the map. A Stephens. Prices start at $250 for the loud speaker unit, but we can get 40% off thru a friend, and I'm tempted to shoot the works and get one, if it can be fitted onto my machine. It's really something.

Haven't seen Ed for couple of months now. Since he went broke, he hasn't been coming here for week ends. Graduation announcement from Cornelia; I haven't been writing to her much; she and I both poor correspondents. Talked to Frances today on the phone but haven't seen her for months. Glad to hear from the Baltimore contingent, give my best to the family. 24 pounds as I recall is very good for 8 1/2 months. From a cup. Can he drink from a glass of cool beer.

In the spring of 1947, Steinbeck fell from a second-story balcony of his home in New York, breaking a kneecap and spraining a foot. He underwent surgery to repair his kneecap and walked first with crutches and then a cane for months afterward, including on his trip to Europe and Russia with photographer Robert Capa.

To Bryant Fitch
November 18, 1947

Dear Bryant:
I was of course delighted to get that good portrait but still more delighted after all this time to be hearing from you. Regrettable you had so little time when you were here last, and the lab was so jamful (as usual). When you come here, let's see if we can't get together; might ring me at Calif. Packing Corp. laboratory if it's during working hours; I'm there on and off.

I saw Jean and Marj the other night, showed them your letter. Jean concerned because she hadn't written; hadn't I guess because confused. As aren't we all. If in my lifetime I come out of all the confusions I've got into, it'll be remarkable. Still there are some clarities, altho not often when they involve obscure relationships. Toni's child, Kay, 12–13 yrs old, just died of brain tumor, and Toni, confused on this and other scores is living now away from me in LA; hard to say how it'll all come out, but I feel alright. Did you get the rather good magazine "What's Doing" that Bruce is publishing? Good job. The review by tj (Toni) on Jn's *Wayward Bus* is perhaps one of the best reviews on John's work that has been published. And I'm interested to see that its being recognized as such; I just finished an account of Steinbeck published in the fall 1947 *Antioch Review* with a statement to that effect. A new edition of *Between Pacific Tides* is in the immediate offing. After hanging fire for so long! Anyway, Stanford sent me advance royalties, so I guess they mean business. Good job, lots of new charts and some color illustr., new section on plankton involving some rather radical ideas—I don't point them out, I just let the evidence lead that way. And a foreword by JS. I have completed two years of a projected 3 year investigation of the Queen Charlotte, W Coast Vanc. Id. region which, when finish, will result in a third book "The Outer Shores" but lord only knows when I can get up there again, ordinary (and our extraordinary) expenses being what they are.

I hope Beth [Fitch] has anyway some fun with writing; I'll be interested to hear how she comes out on that score. Send some stuff to Bruce, who direly needs enough intelligible material to select from. Do say hello to her for me.

>   Bryant Fitch was Jean Ariss's brother.
>   Toni's daughter, Kay, died in early October 1947. With marine biologist Ben Volcani—whom she met in Monterey—Toni left Ricketts and Cannery Row and moved to Los Angeles. The couple were married and moved to Palestine in 1948.
>   Ricketts met Alice Campbell (1922–77 or 1978) in late 1947, and the two were supposedly married in early 1948 in Las Vegas, Nevada. Since Ricketts and Nan were never officially divorced, however, his marriage to Alice was not legal. Alice left Monterey in late 1948.

To John Steinbeck
November 25, 1947

Dear Jn:
Your good and long letter.

Since then a great variety of things has happened so intensely that I am left in some confusion. About the only good of which is that I then have to work out for myself some aspects of human relations which have perhaps universal implications.

It's a pity that the pace of events, or the force of external circumstances push a person into a pattern of action before things are fully worked out in his mind. And that's true especially of human relations—and I wonder now how much I'm speaking symbolically of international activities also. Who I shall be married to is important not only to me, not only to the other person, and the third and perhaps the fourth or fifth involved, but in a sense to all people.— If I have anything to give to all people. And in the "for whom the bell tolls" sense, *all* people have something to give to all.

The relation between Ed and Toni, both of whom are really quite good people I am convinced, has been a tragedy of errors when it wasn't an equal comedy. At the wrong times or in the wrong way, each has had his good thing ready for the other only for it to be reflected back unabsorbed. Now after 6 years, reading over a letter to herself that Toni wrote about us, I realize fully how much she had and how much she wanted to give it to me. And for a long time I knew how much, increasingly, I was able to call back into myself from Jean; which then had no place to go, but which I tried equally hard to put on Toni lately suffering my rebuff of her good thing.

Well, and I suppose Toni must have judged me as unjustly as I judged her, tho it doesn't seem possible. I think if people belong together, their places are with each other. I was upset understandably because Toni went away. But altho I knew, I didn't realize she did so at the suggestion of Evelyn and Auntie Adele, both of them certified psychiatrists. She was in a near nervous breakdown, in a [?] of the emotions—all the more intense probably because in this thing throughout she never showed it externally, never cried once, had no exterior intensity—and in those circumstances understandably she couldn't put out the added energy required by our relationship. Of course it's easy to say that if the relationship had been good, she freely could have rested in it, and wouldn't have had to go away. But it wasn't wholly good. There were lots of bad places, and they caused further strain. Anyway I don't feel very good about my handling of it and we've finally agreed upon what in a conventional relation would amount to a divorce.

After some correspondence about the complete relationship I had entered into with Alice, and the incomplete (on the marital side) relationship she had formed 6 or 8 months back with a young biologist, Toni returned to PG [Pacific Grove] on only 24 hour notice (22 of which had been stolen by the negligent telegraph Co!). I told her that I preferred to live with someone else. And I told her I loved her (Toni)—the extent of which even I hadn't known before. This left her with no alternative but to clear out. I guess I hadn't figured on such great drasticness, or at least not so quickly. Because it had meant now a complete and permanent uprooting. Terrible jolt on Toni; in two months she lost her only child, her husband, her home, her name, even a physical storage place, even a mailing address. In a wonderfully deep and real sense, Toni cannot ever walk alone anymore than Ed Jr can, than Tal can, because I do deeply love her and support her and walk with her. But that's remote help in a shock like this.

The extreme view at one end is that Toni is now embarked on a great adventure. She is going to see the country in a way she'd never see it with me (because I won't go inland) and she's going with a person who in some ways at least, and at present, she prefers to me; and she's going to a place she's always longed for (NY, and you'll probably see her in a few weeks or months). The view at the other extreme is that she came back here recovered from her incipient nervous breakdown, peppy and confident and happy and loving, prepared to take up our relation permanently and in a deeper way than ever before. And she found me in another relationship with a significant person I considered to be more suitable. If I hadn't found such a relationship, and if I had said to her "Toni, I hadn't realized before how much you loved me, how much I loved you (not necessarily how much I'm "in love" with you) and how wonderfully loyal you are" we might have set to work, and perhaps we could have worked out many of the things that had gone wrong, and got into a better relation than we'd ever had. I think I might easily be her most important person, even her No. 1 person; she might have been mine. This whole thing fills me with the most poignant regret; partly perhaps the old saw "the saddest words of tongue or pen . . ." And yet I'm completely sure of my relation with Alice. I suppose it's inevitable there shall be conflicts; and parting with a good friend after seven years is a sad business. Gosh I certainly know some of the things you felt. I've been on the receiving end of this thing before, but never on the delivering end.

We had had only a few great troubles. The sexual thing went a little wrong, that was Toni's Jean-insecurity, but that element is important to me, and I made it worse by complaining and being upset. And Toni became increasingly socially extroverted and gay, so I tired out before she did at parties; she in part substituted social approval for a deep thing in our relation which was lacking. And then she formed a deep friendship with Ben, which had (I know now, but never believed before) no sexual expression but which did have considerable physical affection which Toni'd never been able to give me. And then she went down south, wisely as I know now (but I didn't then) and on advice of physicians. And then I took up with a very good and significant person. And then she wanted to take what little money remained in her account after Kay's sickness (her mad money,) and pay her room and share of food on a trip back to Madison, Wisc. that Ben had to make on his way back to Palestine, as a non-sexual and paying companion, to see the country on a chance she'd never get again and then go on by bus to NY, stay a month, and then come back to Monterey and settle down. And she wrote to ask how I'd feel about it. I said I approved, but only on the basis that our relation was already shot. So then it worked out as I said.

So, jn my friend, that's the whole sad story and I don't feel so good. I guess tho the totality's good. Alice is I'm sure good and significant beyond her 22 years. From her own bent, and training, and from living with a philosophy prof, she's inclined to go right to the essentials of a matter, to regard the superficialities as *only* superficialities. When I first met her I regarded her as confused, now she isn't anymore if even she ever was; I thought she was unreliable, now I know her as the most dependable person I know. So far all the bad things first; and that's a darn good thing. A music-philosophy major at UC, senior, now out of school to live down here with me; the finishing may cause trouble; perhaps in one way no need of finishing, since her interest in scholastics isn't for fame or degree, but for fun and use. I feel completely good on that score, but it certainly breaks my heart to see Toni go; I didn't know how much she meant, how fine she was, how intertwined our roots were. Of course one thought is that I hate to lose contact. As I would with you, or with Xen, or with Ed Jr. When you find the real deep thing, you certainly hate to see it go. Perhaps I have some actual conflict; and with that sense of irrevocableness; not enough time to work it out now.

Well, this has been an intense time. The last stages of Kay's illness could have been pretty bad. We saved her by extreme care; she was perfectly alright. Near as I could see she had very little pain, not much discomfort, no franticness or fear, and very little insecurity if any. Toni put her good licks into that and I can assure you they were good, and they required great skill and discipline. But at what a cost! The gal certainly took a beating. I completely agreed with her about the funeral thing (which was arranged and paid for by Kay's father); in fact it was my affirmation on that score which finally turned the trick. I really don't mind things like that. The *form* of barbarism doesn't so much bother me. For the average run of person there's certainly something to be said for funerals; all I do is to observe and even to some extent to love that. We certainly ought to have more than one word for love. Charity is the best thing we have for pure love, and how sullied *that* is. Say incidentally, in that connection I ran across an interesting illustration. I was trying to work out the business of sexual (or I guess another kind of) promiscuity in a deep relationship. Or in the kind of impurities that kept cropping up in the movie "odd man out"; then I got to thinking about Contrapunctus V (I think it is, may be IV) in the

Art of the Fugue. If you can use that trite term for just a minute, in that if anywhere, the spirit of man soars. Just like the deep love that Dick used to feel for that (miserable) Spike, so deep he just couldn't follow it, I can't follow the soaring in that Contrapunctus; it goes far beyond me. Yet it emerges from the most magnificent discipline! Whatever his temptation to embellish or deviate, Bach follows the fugal discipline unwaveringly and without diverting. He never gets off the track.

Well, it's that time. I'll go cook a big meal for Toni (last for her), Jeanette and Norman [Geshwind] who have given themselves in this so unsparingly, then borrow a car, go out and visit with Alice in a bare room waiting but only superficially lonely and certainly not unsupported (my car's got burnt out bearings, what a weight of trouble now especially) come back and see the gang from HMS [Hopkins Marine Station] who will come over to wish Toni well, thresh out with her still tonight anything yet remaining to be considered that we can bring up in time, and put her on the train for Berkeley in the morning. In another week we'll be physically rested for better or worse.

Ed

Ricketts was mistaken in stating Alice's age as twenty-two; she was actually twenty-five at the time.

To Toni Jackson
Pacific Biological Laboratories
Pacific Grove, California
December 22, 1947

Dear T: The news as I hear it—a to-be-continued. Sally was in finally. Spent the whole evening.

Dec. 24
In the meantime your most welcome letter arrived; but in the meantime let me go on with the sad Raghu-Sally story. They are broken

up. It seems as tho I've heard nothing but separation news for weeks. Sally says it was the morbid possessiveness that translated itself into masochism that finally made the relation impossible. That Raghu got terribly bothered by her riding, that he tended to cosset her about things like the way she ran around on the rocks—danger of falling & breaking her head, that sort of stuff, that finally she was about to tell him that to be jealous of a horse was sort of ridiculous, but wisely she didn't. Or maybe unwisely, I'm not sure. Says that she figured if she went back to India with him she'd never even see a horse again, so she wanted to get in all the riding she could in the meantime. That Raghu took it pretty hard (the separation) having never been in love before, but that it was the third time for her. Then Raghu told me his angle which was mostly of being mystified at the turn of events, that at any rate he was glad that he didn't bring about the separation, and that whatever mysterious reasons she had, they were hers, and had to be respected. He has quit drinking, not even beer (I guess as a punishment to himself) and is winding up his work in this country as rapidly as he can preparatory to going to India right after he gets his PhD, I think he said in June. Based on what Tal said Xen said you told her, I told him Ben had asked you to go back to Palestine with him but I didn't know if you'd do it. Rather suspected you would. Had suspected it right from the start, but it was only a personal opinion. Those things certainly have their huge aura of energy, rather of intense activity, and people on the periphery get involved and interested. And of course people close to, such as I, get intensely concerned, plus or minus. I was once sort of amused in that connection when, when you first came down you regretted that all those intense things had happened in the past before you got here. I thot at the time "Little wormie you'd better be glad they're over, because they can be pretty violent and devastating." And then you and Ben, and subsequently Alice and I, get into just such imbroglios; is that the word I want, intaglios maybe,—a huge complex compelling vortex. That whole thing is most complex and (in one sense) conflicting. I have with Alice an exceedingly good relation of love and confidence and friendship, and trust completely unbroken on both sides—and it's up to me to use what wisdom I've acquired to keep it that way, *not* to spend time in separation for one thing, and to try not to build up a pattern of disagreement (which is slipped back into so easily once it's built up) such as we had to some extent, and which Norman and Jeanette [Geshwind] had pitifully. But in spite of all this—no that isn't it, the things have really no relation. Anyway, I miss you enormously and I love you deeply. The things go right together; certainly what I have for you doesn't diminish my love relation with Alice, it perhaps increases it. And the opposite seems to work also: I'm loving you as deeply and perhaps more clearly than I ever did before. Raghu and I were talking about that; humans suffer mostly from being only human, they suffer from death and from separation, if they were infinite their love could be everywhere, wherever it wanted to, wherever it had to go, and it could be endless and deathless. I should be two people, many people, all of us should; there's the Alice thing, the Toni thing, even still the Jean thing, they're all true. But human limitations and finiteness necessarily cut the proliferations of deeply good things as it does all things. We should be god which we're a part of. That's the enigma of all enigmas: how we can be truly part of God and therefore partake of the whole which is in fact indivisible, and yet have the limitations of being not god. Perhaps only a logical enigma. Times when we truly participate that doesn't apply. But the limitation then is still there: time. The square root of one is impossible to go from a minus to plus; infinity isn't possible, yet we do it in fact all the time, and because we're familiar with it, we forget its mystery. Must as you never think of the fabulousness of the infinite images—the reflections—in a rain washed pavement, the number and the positions of the images being entirely a function of the number and the positions of the observers.

Well you pretty Toni, I've been much interrupted. Steve was just in using the phone here to talk around the country, worried about bonus checks. P'haps I won't get one. Mac has been in and out. Tommy Noice. I am just now working up a wire to send you; just talked to Pat Wall. My love to you dear wormie.

> Raghu Prasad was a graduate student completing his Ph.D. at Hopkins Marine Station.
> Pat Wall owned and operated a modern art gallery in Carmel.

On January 14, 1948, H. F. Preston, of Braun-Knecht-Heimann Company, a supply house in San Francisco, wrote the following to Ricketts:

We have been trying to contact you by telephone for several weeks, but have been unsuccessful, and would like to know just what the status is of the various orders that are still open in our file with you. We also received word from the Sarah Dix Hamlin School, stating there was a long delay in filling their order, and when the material finally arrived it was in very poor condition. [. . .] This is the first time we have ever had any word that you packed material altogether in one container, and we would appreciate advice in this regard.

Ricketts's response reflects his frustration with the situation, which was compounded by other queries and circumstances with which he had to contend.

To Braun-Knecht-Heimann Company
San Francisco, California
January 17, 1948

Dear Mr. Preston:
Your letter of Jan. 14th, plus the present inordinate difficulties of doing business in a small way (and I suppose in a large way equally) make me wonder if I wouldn't do well to get out of this thing completely, go on with my own research, and stress the biochemical assay angle which is a whole lot more profitable than selling biological supplies, and a lot less grief. Only trouble is that too many firms already have got out of this business, and schools are crying not quietly for the biological supplies which are already too hard come by.

When the material suits that method, I have always combined the various items unless specifically requested otherwise. It's much cheaper on containers and reduces the shipment weight.

Just now, when containers are so unavailable, there's even more incentive to packing that way. 2 oz, 8 oz, 2 quart and one gallon wide mouth crew cap jars with closures have been almost unobtainable, and I've devoted a great amount of energy to scouting around, and to procuring and renovating used containers of this sort where (rarely) I could get them. It's worked out, from the standpoint of big business vs small business, that the glass manufacturers are no longer interested in selling a few cases of a given item, or even outright refuse to deal with small users. A large scientific supply house in the city, name of Braun-Knecht-Heimann Co., has proven itself completely unsatisfactory as a source for such items—as unsatisfactory perhaps as an outfit named Pacific Biological Laboratories has proven in the matter of biological supplies—their rates are fantastically increased, delivery is slow, uncertain and incomplete.

I can believe that Sarah Dix Hamlin School was overcritical as a result of delays. I don't blame them a bit. However we've had a lot greater delays here in getting specimens. In many cases, I can't get the boys to collect such items as frogs and crayfish, I have to go out and buy them already preserved at ruinous rates and pay freight or express all the way from back east. And the delay in the container shipment situation has been fantastic, as you should know. Sarah Dix Hamlin pay $1.65 on some miscellaneous items that retailed at only $16.90, for container and carrier charges, and they take minimum express rates. Would a more distant school be willing to pay five or six times that much for separate containers for each large item—gallon jars at 50 or 60¢ each—and high express rates for heavy items and bulky packing? It would certainly be alright with me—if we could get the containers.

I have had the PBL phone unlisted in the Monterey phone book, and I'm thinking now of having my own name unlisted also. Everyone, from oldish feminine shell collectors to casual tourists calls up here, takes up my already limited time to ask me if what they find is animal, vegetable or mineral, asks me when Ricketts and Calvin['s] "Between Pacific Tides" will be republished, or if I can sell them a used copy. Schools call up by long distance wanting sharks or embalmed cats, telling me how badly they need such items, saying price is no object; all I can say is that I haven't got any such and I can't get them. I say the phone is an invention almost as evil as my huge correspondence that I never get a chance to tend to.

Oh I am full of complaints today and your letter is the perfect trigger. Biological specimens are close to unattainable. Where I have to buy them already preserved, rates are high, delivery is slow and uncertain, and even the freight and express rates have increased. And I've been priding myself on keeping all our old prices in force—single exception of frogs. A fine thankless task!

When the specimens are available, I ship 'em. And if containers are decently available, I'll use them and bill the schools for them at cost. Takes twice as long anyway to scrabble around and use makeshifts.

Incidentally, in that connection, what happened to our invoice 12754 dated Dec. 30th, for carrier and container charges on 8 orders previously shipped? This $24.33 item was for cash prepaid at the time of shipment, and should have been paid promptly without discount.

Well, you must forgive me. I suppose this will be only one of many bad-news letters cluttering your desk on a particularly blue Monday. Let's both of us have a cool bottle of beer and forget about the present evils of business.

       Sincerely,

To Toni Jackson
February 4, 1948

Dear T: Maybe a chance for that long letter I've been promising myself. And to answer in detail your several letters which I've replied to shortly.

Jn was as always. He showed up here 4 PM Sunday after calling Saturday from LA in the expectation of getting here not until late the following eve. But there was an all night party, he was a little high. We drank and talked until far into the night, and the following morning he waked up early and started to read. 5 hours sleep for two nights. And then the next night the same, and when I got up early to go to work who should be awake and coughing and smoking but that John. And the night following it was still later. Maybe in the long run I can keep up with him in drinking and unsleeping, but certainly not in any given short period. Now though he's slowing down, says he's sleepy half the time, and seems to be keeping right down to the business of seriously relaxing. We got into quite a profound drunk involving the Heifitz's, with Remo and Nancy drifting in and out. Jn in the meantime razzing Remo good-naturedly for perhaps a full hour about being a phony, then saying to Alice "I like that boy," ending up finally in Betsy going to bed at the lab in Ed Jr's room for a few hrs. to avoid being sick. I collapsing acct no food and too much rum, and Milt, Alice and Jn chasing all over Crml trying to find a Zoe to ease the warmish pants of that John. Then Sam Eskin showed up. It's all very confused with Ritch and Tal and Unka Bob and Cornelia and very genial Toby and Lois and the rest of us in and out of Tal's place. Then one night— fortunately we weren't in on this—Jn and R & T had themselves quite a time. Jn says that Ritch got mad at Tal, at Jn, and even at absent me. I have been out as much as possible because as soon as it got noised around that Jn was in town, working up for a movie etc, all the movie mad people of the region and the celebrity hunters, not being able to find Jn, grabbed onto me. And I can stand just so much of that social extroversion. Jn himself for that matter can't. He poops out about the time I do. And Alice can stand less than any of us. So in spite of all the publicity we've managed a few completely quiet nights, and none with any strangers. Last night Jn came over alone for dinner, no one else showed up, we had a quiet evening of Jn reading, Alice playing the phonog. and I working on ideas in my notebook. Of the very greatest things *The Art of the Fugue*, Don Giovanni, Goethe's *Faust*, the Beethoven *Quartette No. 16*, and *Finnegans Wake*. Now I know the *Wake* is the greatest book I've ever come in contact with, greater even than *Faust*. I got excited to the point of translating again the last few lines of the 2nd part.

And in that connection, do you have any recollections on my pretty German *Faust*—the one that Evelyn gave me. And on the MacI *Faust*—the one in English and German that I read you that pretty night so long ago? Been gone sometime. Last time I remember the MacI—or perhaps the both of them—was in connection with your talk in Carmel when that woman challenged you, and we dug out the MacI and some originals, worked out "calls a spade a spade" for the line "Wer dar das Kind beim rechten Namen nennen . . ." and you wrote to the woman. Did you take the volumes to Crml, or perhaps leave them in a restaurant somewhere. Don't bother replying unless you recollect something. If no reply, I'll assume there's no lead there.

So then anyway I considered those 5 great things, and several lesser. Don Giovanni, Contrapunctus XIX of *Art of the Fugue,* and the Beethoven late quartettes finally break, and go down into noble tragedy. The statue comes for Don Giovanni who fears for a moment, then recovers himself, carries on this magnificent evil, refuses to be saved by the prayers of Dona Elvira (is it?) and allows himself to be dragged down to hell. Contrapunctus XIX goes into that great thing which I think of as beyond life (and death), it speaks out of it, but no Bach could finish it, and he dies with it uncompleted on that magnificent shrieking high note suddenly, as Biggs, bless his heart, plays it on the Harvard Baroque organ. The lesser Beethoven glimpses that beyond, he speaks out of it for a moment and dies. All of these show magnificent tragedy, but tragedy nevertheless. But then consider the affirmation of Goethe and the *Wake*. It's funny, of the two very greatest—for me the greatest in the world; one—the *Fugue*, is "negative," the other, *Finnegans Wake*, is affirmation.

The last of Pt. I *Faust* is obvious, Mephistopheles saying of Margurite "she is lost" and the chorus of angels saying "she is redeemed." Literal translation of the last of Pt. II "All passing-ness Is only a likeness The inadequate. Here becomes actuality. The indescribable. Here is it done. The eternal womanness. Draws us up." Or actually: "All transient things are only symbols (likenesses, similarities, shadows). The potential here becomes fulfillment. The unthinkable is here achieved. The eternal principle of femininity leads us on." The original is deep and natural; translations are inadequate; the one best known is terrible. "All of mere transient date, as symbol showeth. Here the inadequate to fullness groweth. Here the ineffable. Wrought is in love. The ever-womanly draws us above" even tho it does preserve the meter of the original. The eternal principle of femininity could be expressed by Eros, the symbol for relation, relatedness, which implies an apperception of the whole, which is in turn related to a sense of proportion. But the *Wake* is best.

Coming. Far!
End Here.
Us Then.
Finn, again.
Take.
Bussoftlhee, mememormee!
Till thousendsthee.
Lps.
The keys to.
Given.
A way a lone a last a loved a long

Sometimes when I read that even to myself it seems that humans aren't great enough to bear the visions they conceive. Softly.Thee.He.Memory.More me. Thou ends thee. thousands end. thousands end in thee. he. And that little lisping sound when current meets wave. The sound that scared Jn and me in the gulf, whistle wind on water that turned to hissing small waves on the shore. Take Finn again and again. Take the keys. And that magnificent fine final affirmation: Given.

Well I got so elevated, and on that yesterday I was so tired from no sleep, and I had the curious sadness of knowing then that you were off to Palestine, and [an] allergy I hadn't had for a long time, that it all piled up to I guess the worst choking spell I ever had. A person's throat (which is a part of the person) closes up so it won't give him air; one part threatens suicide to the whole; even a happy whole. I got that sadness of things that aren't any more; as when I talk to Kay in the still same back room (I've had no time to change it around). Es war nicht gewesen sein. It was not to be. And of course: OK. Because everything can't be. All potential can't become reality. You've got to select. But it makes you sad.

> Steinbeck visited Ricketts in Cannery Row during the first few days of February 1948. The two discussed their impending trip to the Queen Charlotte Islands off the coast of British Columbia that July. Steinbeck was to finance the trip and hoped Gwyn would join them. This was to be the last time Steinbeck would see Ricketts.

To Joel W. Hedgpeth
February 17, 1948

Dear Joel:
Prompt is I'm. II [edition of] BPT is not yet out and I don't think it can be for many months. Words cannot describe Stanford Press but a letter that Jn and I just wrote them perhaps can get across the idea. Among other things we requested one copy of their fascinating new publication: "The Internal Combustion Engine, Will It Work." We said also "Science progresses, Stanford Press does not." I think it will either bring promptness or anger. But since they sent advance royalties I think they'll go ahead.

I expect to be in the Queen Charlottes again this summer, probably gone when you come thru; person to person call will tell if you're where phone is easy. I will expect to finish up the survey for a third book "The Outer Shores" then, and will start right in on it, heaven permitting, this fall. I think Stanford will be glad to publish, eventually, if they aren't too mad. In that connection, keep in mind the little pycns [pycnogonids]; I want to get all that as complete as possible. Tentative identifications even will be welcome.

Texas sounds to me like a bad bad place. The political situation especially—a focus of all that's racist, fascist, nationalist or "I'm-better'n-you-by-god" featuring the Texas hill billies with Ma so and so and her juiceharp.

Coll Pac [College of the Pacific] seem to be doing a fine thing with their Dillon Beach [marine station] project; Noble was certainly on the job to get it. I think you'll have a lot of fun there, and I will hope we can get to see you in that connection.

Hastily
Ed

In the final weeks of his life, Ricketts continued his plans for a trip to the Queen Charlottes in the summer of 1948. He and Alice were to be joined by Steinbeck, whose marriage to Gwyn had become turbulent. This letter to Campbell, the latest available, is likely one of the last pieces of writing he composed. Edward F. Ricketts Jr. believes, however, that later letters may exist and that "it's possible that we'll never know for certain the last thing he wrote" (Interview, August 31, 2000).

To Joseph Campbell
Pacific Biological Laboratories
Pacific Grove, California
April 26, 1948

Dear Joe:
Plus the regular once-a-year letter need, one or two added things, two or three.

Best, your fine book just received; medieval illustrations I wish there were more. Sometimes I get a strong sense of contact with that time. I am glad to have Gawaine and the green knight; the Mallory is different? But this is original in the sense the Mabinogion is. You must be doing a great lot of work. Last year's book too. A Hindu friend of ours, no I should say an Indian friend, a good zoologist has it loaned out now.

Fellow who thinks in little but the terms of math, chemistry, physics; queer business for an oriental.

Don't do too much; but then probably you will (what with teaching). As I do. Less leisure. Less proverbial inviting of the soul. And my annual sardine report; copy of which will be going forward to you.

Is it this year you'll be coming thru? Not until August; we'll be back then from the north. I do think in all seriousness those things are important; if a person can't do them from standpoint of finances then I suppose that has to be accepted; if from standpoint of time, may be arrangeable anyway without too much sacrifice. I think: by being efficient, but then I don't seem to be getting that way very fast.

We'll be going again to the Queen Charlottes end of May. John is coming up there for part of July. I have turned over to him verbatim transcriptions of my two summer's notes; then he'll have his own and mine for the coming trip. Should be able to construct quite a book out of them; he'll have his journal done I fear far before my scientific part's complete. It should be a smaller *Sea of Cortez*. "The Outer Shores." I'll send you one. The II Ed [of] BP Tides still isn't rolling. Galley proof two months back. There'll be a separate of the plankton essay which I'll send. I think it's very good.

Jn is financing this one, and I'll pay him back my share out of royalties if any, if enough. Altho as you say there's not much chance. I will hope to work up a request for subsidy at your Bollinger's, similar to Guggenheim, for 1949 or 1950. (The one year's delay would have been necessary anyway, [on] account of Kay. Toni and I put everything we had into that, personally, financially, in a work sense; she of course far more than I, and deserves every credit conceivable on that score at least). I'll probably apply again to Guggenheim also; do no harm. I can't imagine why I missed it then, it seemed as tho I were all set.

I am getting ideas on ecology more and more; no end in sight immediately. Dr Stephenson (U. Wales, U. So. Afr.) was here for a while;

wonderful fellow; one of the greatest ecologists. I envy very much his ability and efficiency. Works here two or three months, knows many things about the local animals I've never learned, some of them important. Good man personally, even amusing; a gentle radical in any field he touches; amusing pen and ink sketches of his wife or of some equally charming naked woman, swimming among the sea animals, must bother some of the stodgy editors to print them.

Toni's by this time in Palestine. I seem to be the only one who anticipated it; and that I suspected six months ago before, apparently, it was even in her own mind. (Tal so upset she lost fifty million dollars to me on it.) She isn't much for writing. But then she never was, even when with me, and I suppose now I'll hear not at all. A funny girl; careless almost to the point of criminal carelessness, but with intent loyalty I guess almost completely dependable; as I think for ever she can depend on my friendship [and] loyalty however angry I get at her circumstantially. Well, time to go; come along out to PG again. Ritch and Tal are still holding forth with more children than ever. Everything else is pretty much the same. My best to Jean.

Ed

Works Cited

Allee, W. C. *Animal Aggregations: A Study in General Sociology.* Chicago: U of Chicago P, 1931.
Astro, Richard. *John Steinbeck and Edward F. Ricketts: The Shaping of a Novelist.* Minneapolis: Minnesota UP, 1973.
———. Introduction. *The Log from the* Sea of Cortez. By John Steinbeck. New York: Penguin, 1995. vii–xxiii.
Benson, Jackson. *The True Adventures of John Steinbeck, Writer.* New York: Viking, 1984.
Berggruen, Heinz. Letter to Edward F. Ricketts. Apr. 23, 1942. Edward F. Ricketts Papers. M0291. Special Collections, Stanford University.
———. Letter to Edward F. Ricketts. June 1942. Edward F. Ricketts Papers. M0291. Special Collections, Stanford University.
———. Letter to Edward F. Ricketts. Aug. 5, 1942. Edward F. Ricketts Papers. M0291. Special Collections, Stanford University.
Campbell, Albert. Rev. of *Between Pacific Tides,* by Edward F. Ricketts. *Monterey Peninsula Herald,* Nov. 2, 1939: n.p.
Campbell, Joseph. Letter to Edward F. Ricketts. Sept. 14, 1939. Edward F. Ricketts Papers. M0291. Special Collections, Stanford University.
———. Letter to Edward F. Ricketts. Dec. 10, 1939. Edward F. Ricketts Papers. M0291. Special Collections, Stanford University.
———. Letter to Edward F. Ricketts. May 19, 1940. Edward F. Ricketts Papers. M0291. Special Collections, Stanford University.
Carberry, Edward. Rev. of *The Wayward Bus,* by John Steinbeck. *Cincinnati Post,* Feb. 15, 1947.
Clay, David M. Letter to Edward F. Ricketts. Aug. 20, 1942. Edward F. Ricketts Papers. M0291. Special Collections, Stanford University.
Cox, Martha Heasley. "Interview with Peter Stackpole, Photographer." *The Steinbeck Newsletter* 9.1 (1995): 19–23.
de Kruif, Paul. Letter to Edward F. Ricketts. Oct. 20, 1939. Edward F. Ricketts Papers. M0291. Special Collections, Stanford University.
DeMott, Robert. Introduction. *To a God Unknown.* By John Steinbeck. New York: Penguin, 1995.

———. *Working Days: The Journals of* The Grapes of Wrath. New York: Viking, 1989.

Englert, Peter A. J. "Education and Environmental Scientists: Should We Listen to Steinbeck and Ricketts's Comments?" In *Steinbeck and the Environment*. Ed. Susan F. Beegel, Susan Shillinglaw, and Wesley N. Tiffney Jr. Tuscaloosa: U Alabama P, 1997. 176–93.

Fadiman, Clifton. Rev. of *Sea of Cortez*, by Edward F. Ricketts and John Steinbeck. *New Yorker*, Dec. 6, 1941.

Gibbon, Bob. Letter to Edward F. Ricketts. Jan. 15, 1946. Edward F. Ricketts Papers. M0291. Special Collections, Stanford University.

Hedgpeth, Joel. *The Outer Shores*. 2 vols. Eureka: Mad River Press, 1978.

———. "Philosophy on Cannery Row." In *Steinbeck: The Man and His Work*. Ed. Richard Astro and Tetsumaro Hayashi. Corvallis: Oregon State UP, 1971. 89–128.

Jackson (Volcani), Toni. Personal interview. Dec. 7, 2000.

———. "Peter Ferrante: Civic Citation." *What's Doing* 1.6 (1946): 24, 36–37.

Kelley, James C. "John Steinbeck and Ed Ricketts: Understanding Life in the Great Tide Pool." In *Steinbeck and the Environment*. Ed. Susan F. Beegel, Susan Shillinglaw, and Wesley N. Tiffney Jr. Tuscaloosa: U Alabama P, 1997. 27–42.

Larsen, Stephen, and Robin. *A Fire in the Mind: The Life of Joseph Campbell*. New York: Doubleday, 1991.

Macaulay, E. L. Letter to Edward F. Ricketts. Oct. 14, 1939. Edward F. Ricketts Papers. M0291. Special Collections, Stanford University.

McConkie, J. E. Letter to Edward F. Ricketts. June 29, 1943. Edward F. Ricketts Papers. M0291. Special Collections, Stanford University.

Preston, H. F. Letter to Edward F. Ricketts. Jan. 14, 1948. Edward F. Ricketts Papers. M0291. Special Collections, Stanford University.

Price, Bob. Letter to Edward F. Ricketts. Aug. 27, 1942. Edward F. Ricketts Papers. M0291. Special Collections, Stanford University.

Railsback, Brian E. *Parallel Expeditions: Charles Darwin and the Art of John Steinbeck*. Moscow: U of Idaho P, 1995.

Ricketts, Anna Maker. "Recollections." Ts. 1984. Center for Steinbeck Studies, San José State University.

Ricketts, Edward F. *Between Pacific Tides*. Rev. ed. Stanford: Stanford UP, 1948.

———. "EFR Essay No. 2." Ts. Edward F. Ricketts Papers. M0291. Special Collections, Stanford University.

———. "New Series Notebook No. 1." Ms. Edward F. Ricketts Papers. M0291. Special Collections, Stanford University.

———. "New Series Notebook No. 2." Ms. Edward F. Ricketts Papers. M0291. Special Collections, Stanford University.

———. "New Series Notebook Number 3: Started Late June 1944 Including Outside Coast Vancouver Island Expedition June July 1945." Ms. Edward F. Ricketts Papers. M0291. Special Collections, Stanford University.

———. "New Series Note Book Number Four Started Sept 1945, After First Trip to West Coast of Vancouver Island." Ms. Edward F. Ricketts Papers. M0291. Special Collections, Stanford University.

———. *Pacific Biological Laboratories 1925 Catalogue*. N.p.: n.p., 1925.

———. "[Palao 1942]." Ts. Courtesy of Edward F. Ricketts Jr.

———. "Scientist Writes About Art and Defines 'Abstraction.'" *Monterey Peninsula Herald*, Oct. 31, 1941, 10.

———. "Second 1940 Mexico Trip Notebook One." Ms. Edward F. Ricketts Papers. M0291. Special Collections, Stanford University.

———. "Suggested Outline for Hndbk of Marine Invert. of SF Bay Area." Ts. Edward F. Ricketts Papers. M0291. Special Collections, Stanford University.

———. "Synonymy of the Palao Islands." Ts. Courtesy of Edward Ricketts Jr.

———. "Thoughts on My First (Substantial) Taking Over of the Responsibility of a Parent." Ts. Edward F. Ricketts Papers. M0291. Special Collections, Stanford University.

———. "Vagabonding in Dixie." *Travel* (June 1925): 16–18, 44, 48.

———. "Verbatim Transcription of Notes of Gulf of California Trip March April 1940." Ts. Center for Steinbeck Studies, San José State University.

———. "Worth Remembering Concepts Arrived at or Considered during 1932 BC Alaska Trip." Ts. Edward F. Ricketts Papers. M0291. Special Collections, Stanford University.

———. "[1946 Sardine Report]." Ts. Edward F. Ricketts Papers. M0291. Special Collections, Stanford University.

———. "1946–47 Guggenheim." Ts. Edward F. Ricketts Papers. M0291. Special Collections, Stanford University.

———. "[1947 Sardine Report]." Ts. Edward F. Ricketts Papers. M0291. Special Collections, Stanford University.

Ricketts, Edward F., Jr. E-mail to the author. July 13, 2000.

———. E-mail to the author. Aug. 11, 2000.

———. E-mail to the author. Sept. 5, 2000.
———. E-mail to the author. Oct. 3, 2000.
———. E-mail to the author. Oct. 11, 2000.
———. Personal interview. June 28, 2000.
———. Personal interview. Aug. 31, 2000.
———. Personal interview. May 11, 2001.
———. Personal interview. July 12, 2001.
———. Personal interview. Oct. 12, 2001.
Ricketts, Nancy. Letter to the author. Jan. 17, 2002.
Ritter, William Emerson, and Edna W. Baily. "The Organismal Conception: Its Place in Science and Its Bearing on Philosophy." *University of California Publications in Zoology* 31 (1931).
Steinbeck, Elaine, and Robert Wallsten, eds. *Steinbeck: A Life in Letters*. New York: Viking Penguin, 1975.
Steinbeck, John. "About Ed Ricketts." In *The Log from the* Sea of Cortez. New York: Penguin, 1995. 225–74.
———. *Cannery Row*. New York: Penguin, 1995.
———. *The Grapes of Wrath*. New York: Penguin, 1992.
———. *The Long Valley*. New York: Penguin, 1995.
———. *The Moon Is Down*. New York: Penguin, 1995.
———. *To a God Unknown*. New York: Penguin, 1995.
———. *Tortilla Flat*. New York: Penguin, 1997.
Steinbeck, John, and Edward F. Ricketts. *Sea of Cortez: A Leisurely Journal of Travel and Research*. New York: Viking, 1941.
Strong, Frances Ricketts. Notebook. Ts. Courtesy of Edward F. Ricketts Jr.

Index

References to letters are printed in boldface

Albee, Dick and Jan, 14, **20–22,** 25, **28–30,** 32, 104, 148, 195
Allee, Warder Clyde, xix–xx, xxviii, 1
Ariss, Bruce and Jean, 24–26, 256
art (appreciation of, discussion of), xxxvi, 11–12, 18–20, 46–47, 69, 121
Art of the Fugue, The, xiii, 204, 261, 267–68. *See also* music
Astro, Richard, xi, xviii, xx, xxiii, xxiv

Bach, Johann Sebastian, xiii, xxxvi, 24, 36, 51, 204, 261, 267–68. *See also* music
Beethoven, Ludwig Van, 33, 204–5, 267–68. *See also* music
Belloc, Hilary, 7, 24, 26, 56, 107, 161, 195, 216
Berggruen, Heinz, xxxiv, **145–46, 149–53, 157–59**
Between Pacific Tides, xii, xx, xxx, xlii, l, liii, 165, 176; publication of, 18, 24, 34, 36; response to xxxii, 113, 115; revision of, xxxix, 123, 125, 132, 172, 211, 219, 234–35, 252, 254, 256, 265, 270, 272
Bicknell, Harold "Gabe," 16–17, 22, **37–38**

"Black Marigolds," xi, xxxv, 197. *See also* poetry
Bogard, V. E., **1–3, 22–23, 26–27,** 40, 137, 139
Braun-Knecht-Heimann Co., **264–66**
breaking through, xxxv, xlv, 18–19, 115, 152, 235
"Breaking Through, The Philosophy of" (essay), xiii, xli–xlii, 14, 44, 50, 68, 115
Budd, Paul, 14, 20–21

Cage, John, xxxvi, 33, 34, 44, 81–84, 120–21, 194
Cage, Xenia, xxviii, xxxvi, 33, 34, 44, **46–47,** 89, 120, **185–88,** 194, **196–99, 203–6, 212–15,** 245, 252
Calvin, Jack and Sasha, xxiii, xxvii, xxx, 3, 35, 44, 159
Campbell, Alice. *See* Ricketts, Alice Campbell
Campbell, Joseph, xiii, **43–46, 52–54, 68–70, 135–36, 159–61,** 188, 191, **206–7,** 238, 240, **245–47,** 248, **250–54, 271–73;** affair with Carol Steinbeck, xxvi–xxvii; living in Monterey, xxxvi; Sitka Alaska trip, xxvii–xxviii
Cannery Row, xii, xiv, xlviii, il, li,

liii, 1, 3, 21, 32, 47, 140, 162, 163, 167, 206, 207, 209, 210
Cannery Row, xiv, xviii–xix, xxiii, xlii, xlvii–il, 16, 22, 37, 208, 209, 210; film, 216–17
Carlson, Norm, **141–42**, 143, 145
Cohee, John and Alice, **14**, 24, 73, 195
Conger, Gwen. *See* Steinbeck, Gwen
conservation (ecological), xii, xxii, xlvii–xlviii, 243–45, 246–47
Covici, Dorothy, **55**
Covici, Pascal, xlv, **54–55**, 73, **119, 247–48**

Dekker, Bill, 27, 29, 39
De Kruif, Paul, xxxiii, 45, **50–51**, 53

Ecological holism, xiii, xix–xx, xl–xli, 240
Emblem, Don, **183–85**
Enea, Sparky, **155–57, 162–64, 166–67, 173**

Farmer, Betty, **235–37**
Faust, xiii, 4, 108, 145–46, 161, 233, 248, 267–69
Finnegans Wake, xiii, 46, 54, 136, 207, 245, 248, 267–68, 269
Fish and Game, Division of, **47–50**
Fitch, Bryant, **255–56**
Fitzgerald, James, xxxi, 1, 12, 14, 17, **24–46,** 56, 73, 161, 164, 186, **193–96**
Fitzgerald, Peg, 24, **56,** 73, **105–8,** 186
Flanders, Austin and Hazel, **1, 12–13,** 40
Flanders, Howard and Emma, **10–12, 31–32, 36**

Galigher, Albert E., xx, xxi, xxii, 86
German, Department of, UC Berkeley, **15**
Gibbon, Bob, **220–22**
Gislen, Torsten, xxxii, 1, **17–18, 33–35, 176–78**
Goethe, Johann Wolfgang von. *See Faust*
Graham, Barb and Ellwood, xiii, 111, 112, 123–25, **154–55,** 168, 187, 216, 239, 253
Gramophone Shop, **52, 248–49,** 254. *See also* music
Grampus, xxviii, 44, 88, 159
Guggenheim fellowship, l–li, 35, 226, 239, 246, 248, 250, 253, 272

Harcourt, Brace, and Co., xv, **164–65**
Hedgpeth, Joel, xx, xxiii, xxviii, xxix, 1, **125–26, 131–33,** 234, **270–71**
Hemingway, Ernest, 197; *To Have and Have Not*, 32, 101–2
Holism, xx, xxv–xxvi, xxxiii–xxxiv, xl, 69–70

Ingersoll-Waterbury Co., xviii, **178–79**

Jackson, Kay, xliv, xlviii, 79, 98, 123, 168, 182, 190, 199, 203, 213, 223, 225, 228–29, 235–37, 238, 240, 245, 249–50, 254; death of, il, l, li, 256, 260, 270, 272
Jackson, Toni (Volcani), xliv, xlvi, xlviii, il, l, li, **76–79, 81–84,** 96, 97, 98, **100–103,** 108, 113, 115, 118, 119, 123, 128–29, 141, 145,

148, 152–53, 160, 166, 168–69, 170, 177, 182, 183, 190, 192, 193, 199, 214–15, 223, 224–25, 235–37, 238, 239–40, 245, 249–50, 252, 254, 256, 257–61, **261–63, 266–70,** 272–73
Japan, Mandated Islands. *See* Palao
Jeffers, Robinson, xxxv, 4, 114, 120, 253. *See also* poetry
Jung, Carl, xxviii, xxxiii, 45, 152. *See also* philosophy

Kashevaroff, Xenia. *See* Cage, Xenia
Kline, Herb and Rosa, xliii, **57–58,** 59, **70–71,** 73, **75–76**

Lannestock, Gustaf, 25–26, 27, 33, 100
Lao Tse, xiii, 234. *See also* Taoism
Lloyd, Frank and Marge, 7, 20, 56, 107, 182, 186, 194
Lovejoy, Ritch and Tal, xxix, xxx, xliii, xliv, 33, 35, 44, 56, **59–68,** 74–75, 76, 80, 90, 100, 104, 111, 123, 161, 185–86, 190, 194, 196–97, 199–200, 202, 204, 238, **241–45,** 258, 267, 273

McConkie, J. E., **180, 211–12**
Maker, Anna Barbara. *See* Ricketts, Anna
Milestone, Lewis, 120, 168, 239, 240
Miller, Henry, xiii, xxxvi, xliv–xlv, **113–16,** 120, 135, 183, 184
Monterey County Hospital, **16–17**
Monterey Peninsula Herald, xlvii, **217–18**
Museum of Modern Art (NY), **19–20.** *See also* art
music (appreciation of, discussion of), xiii, xxxv, 28–29, 51–52, 84, 103–4, 121, 204–6, 212–14, 248–49, 254–55, 267–68

Naval Procurement Bureau, **161–62**
Neiman, Gilbert and Margaret, xliv, 113, 115, **119–122**
nonteleological thinking, xxv, xxvi, xxxiii, xxxv, xxxvi, xliii, li, 44, 58
"Non-teleological Thinking" (essay), xxxiii–xxxiv, xli, xlv, 14, 44, 50, 52–53, 58, 100, 114, 120, 134, 135

Ott, Evelyn, 21–22, 195, 257
"Outer Shores," xxix, xlvii, 1, li–lii, liii, 126, 207, 227–28, 236–37, 251, 256, 270, 272

Pacific Biological Laboratories, xii, xxi–xxii, xxiii, xxxviii, xlv–xlvii, liii, 3; business matters, 22–24, 26–27, 34, 37–38, 47–50, 139–40, 173, 177, 223, 251, 264–66; fire, xxxi–xxxii, 1, 8–11, 12–13, 17, 43; Steinbeck as vice president of, 40–43
Palao, xlvi–xlvii, 141–42, 142–44, 145, 171–72, 187, 195, 201, 207, 208–9, 216
philosophy (appreciation of, discussion of), xiii, xxxiii–iv, xxxv, 53–54, 151–52, 160–61, 228–31
plankton, xii, xxii, xlviii, 211–12, 219, 224, 226, 241–42, 251, 252–53, 256. *See also* conservation
poetry (appreciation of, discussion of), xiii, xvii, xxxiv–xxxv, 4–5, 10–11, 14–16, 81–83, 119, 120–22, 183–85, 197
Price, Bob, 159, **169–73**

Queen Charlotte Islands, Canada.
See "Outer Shores"

Ricketts, Abbott, xv, xxi, 17, 43
Ricketts, Alice Beverly Flanders, xv,
 xxi, 24, 95, 231–33, 238
Ricketts, Alice Campbell, li, liii,
 256, 258–61, 262–63, 267, 271
Ricketts, Anna Barbara Maker
 (Nan), xxi, xxii, xvii, xxxi, xlv,
 li, 3, 17, 24, 30, 34, 43, 58, **72–
 75**, 87, 103, 107, **136–37, 174–
 75**, 177, 187, **188–89**, 198, 203,
 226, 256
Ricketts, Cornelia, 30, 59, **71–72**,
 87, 105, 107, **112, 168–69**, 177,
 182, 187, 189, 194, 199, 202–3,
 238, 250, 255
Ricketts, Edward F., Jr., xxi, xxx,
 xxxix, xlviii, li, 16–17, 22, 23,
 24, 30–31, 33, 34, 39, 43, 51,
 73, 81, 87, 95, 98, 103, 104–5,
 107, 111, 112, 137, 156, 162, 170,
 175, 177, **181–82**, 187, **190–91**,
 194, **200–203, 208–11**, 212,
 215–17, 223–24, 226, 228, 233,
 234, 238, 240, 250, 251, 258, 271
Ricketts, Frances. See Strong,
 Frances Ricketts
Ricketts, Nancy Jane, 24, 30, 58,
 71–72, 87, **103–5**, 107, **110–11**,
 138–41, 174, 187, **189–90**, 194,
 202, 208, 210, **223–24**, 226,
 232, **237–39**, 250, **254–55**
Ricketts, Thayer and Evelyn, xv, **8–
 10**, 232
Ritter, William Emerson, xxiii–
 xxiv, xxv

sardines, xii, xlvii–xlviii, 191, 211–
 12, 226, 227, 237, 241–45, 246,
 250, 272. See also conservation

Scardigli, Remo, 26, 33, 39, 139,
 192–93, 199–200, 267
Scardigli, Virginia, xliv, **32–33, 38–
 39**, 100, 171
Sea of Cortez, xii, xiii, xxxvi, xxxviii–
 xlii, xliii–xlvi, liii, 54, 56, 57, 58,
 68, 72–73, 80, 93, 100, 105,
 114, 116–18, 120, 124–25, 131,
 133, 135, 164, 176, 206, 209, 272
Shan Kar, Uday, 25, 26, 28, 29. See
 also music
shark liver oil, xiii, xxii, 4, **5–8**, 27,
 54, 55, 105, 191, 194, 220–22
Solomons, Theodore Seixas, **148–
 49, 249–50**
Spanish, Department of, UC
 Berkeley, **15–16**
"Spiritual Morphology of Poetry,
 The," xxxiv, 14, 44, 69, 120. See
 also poetry
Steinbeck, Carol, xxii–xxiii, xxvi,
 xxxvii, xl, xliii, xlvi, 18, 21,
 24, 33, 44, 56, 59, 77–79, 81,
 109, 111, 122–23, 126, 137, 139,
 194, 202
Steinbeck, Gwen, xlvi, li, 77, 79,
 80–81, 108–110, 111, **122–24**,
 126, 139, 155, 190, 194, 196,
 204, 209, 212, 228, 254, 270
Steinbeck, John, xi, xiv, xxxvii,
 lii, **40–43**, 59–61, **116–18**,
 142–44, 147, 182–83, 189,
 224–37, 239–40, 257–61,
 266–67; "About Ed Ricketts,"
 xi, xiii, xvii–viii, xlvii, lii–liii, 1;
 and Carol Steinbeck, xxii–xxiii,
 xxvi–xxvii, xlvi, 18, 77–79, 109–
 10, 126, 127–31, 137, 139; *Forgotten Village, The,* xliii–xliv, 58, 73,
 76; *Grapes of Wrath, The,* xxv,
 xxxvii, 67, 75, 79; and Gwyn
 Steinbeck, xlvi, 79, 81, 108–10,

126, 127–31, 139, 155, 190, 194, 203–4, 212, 215, 254–55; *Log from the* Sea of Cortez, *The,* xiii, liii; Mexico City trip, 59–61, 67, 70–71, 74; *Of Mice and Men,* xxxiv, 17–18, 39; "Outer Shores," lii–liii, 270, 271, 272; phalanx theory, xxiii–xxv; "San Francisco Bay Area Handbook," xxxviii–xxxix, 54, 56; *Wayward Bus, The,* 248, 250, 256. See also *Cannery Row* and *Sea of Cortez*

Stevens, Jewel, **133–35**

Street, Webster (Toby), 41–43, 111, 126, 127, 129–30, 166, 187, 194, 202, 210

Strong, Frances Ricketts, xv, xvi, xix, xxii, xxiii, 1, 2, 8, 10, 21–22, 24, 30, 40, 135, 181, 190, 206, 208, 232–33, 238, 255

Strong, Fred, xxii, 1, 21, 24, 30, 33, 96, 111, 176, 194, 238

Taoism, xiii, xxxiii–xxxiv, xliv, 4, 53, 82, 152. *See also* philosophy

Volcani, Ben, li, 256, 259, 262

Washington, University of, **219**
Western Flyer, xxxix, xl, xliii, 57, 93, 157
What's Doing, 26, 236, 240, 253, 256
Whitman, Walt, xiii, xvii, xxxv, 4, 10, 114, 185. *See also* poetry
Wood, Charles Erskine Scott, **3–5, 5–8**
Woods, Flora, 28–29, 47, 163, 206